Chemical Bonds

AN INTRODUCTION TO ATOMIC AND MOLECULAR STRUCTURE

HARRY B. GRAY
California Institute of Technology

Chemical Bonds

AN INTRODUCTION TO ATOMIC AND MOLECULAR STRUCTURE

University Science Books
Mill Valley, California

University Science Books
20 Edgehill Road
Mill Valley, CA 94941
Fax: (415) 332-5393

Library of Congress Catalog Number: 94-61186

ISBN: 0-935702-35-0

Printed in the United States of America
10 9 8 7 6 5 4 3 2 1

Preface

With this edition, University Science Books becomes the publisher of *Chemical Bonds: An Introduction to Atomic and Molecular Structure*. We thank the publisher, Bruce Armbruster, for his interest in chemistry and chemical education.

This book is intended to provide a reasonably complete introduction to atomic and molecular structure and bonding for science students. Parts of the book are revised and expanded versions of appropriate sections from *Basic Principles of Chemistry,* which I coauthored with Gilbert P. Haight, Jr. The basic approach of using illustrations profusely in presenting concepts has been retained in this monograph.

The material on molecular structure is organized roughly in order of molecular size, proceeding from diatomic molecules in Chapter 3 to the "infinitely large" atomic clusters in Chapter 6, which deals with the structures of solids. Although Chapter 3 is loaded with "teaching molecules" (simple molecules observed only at high temperatures and low pressures), the emphasis in the rest of the book is on "real molecules." Each chapter concludes with a large selection of questions and problems.

I should like to thank especially Dr. James L. Hall and Mr. John R. Nelson, who edited the manuscript and made helpful critical comments on each of the several drafts of the manuscript. I am also grateful to Mr. Michael Bertolucci, who read the final draft and offered additional comments and verified the latest values of physical data from several literature sources. At this point an author usually says he must, of course, take responsibility for any remaining errors and impossibly incoherent sections. With all due respect, I should like to share this responsibility with Jim, John, and Mike. So, students and other readers, write and let us know what you think of the book.

Harry B. Gray

Contents

1

Atomic Structure

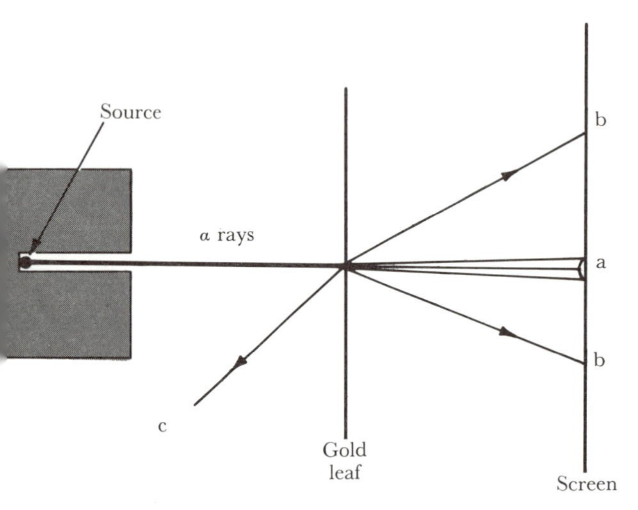

1–1
The experimental arrangement for Rutherford's measurement of the scattering of α particles by very thin gold foils. The source of the α particles was radioactive radium, encased in a lead block that protects the surroundings from radiation and confines the α particles to a beam. The gold foil used was about 6×10^{-5} cm thick. Most of the α particles passed through the gold leaf with little or no deflection, a. A few were deflected at wide angles, b, and occasionally a particle rebounded from the foil, c, and was detected by a screen or counter placed on the same side of the foil as the source.

The concept that molecules consist of atoms bonded together in definite patterns was well established by 1860. Shortly thereafter, the recognition that the bonding properties of the elements are periodic led to widespread speculation concerning the internal structure of atoms themselves. The first real breakthrough in the formulation of atomic structural models followed the discovery, about 1900, that atoms contain electrically charged particles—negative electrons and positive protons. From charge-to-mass ratio measurements J. J. Thomson realized that most of the mass of the atom could be accounted for by the positive (proton) portion. He proposed a jellylike atom with the small, negative electrons imbedded in the relatively large proton mass. But in 1906–1909, a series of experiments directed by Ernest Rutherford, a New Zealand physicist working in Manchester, England, provided an entirely different picture of the atom.

1–1 RUTHERFORD'S EXPERIMENTS AND A MODEL FOR ATOMIC STRUCTURE

In 1906, Rutherford found that when a thin sheet of metal foil is bombarded with alpha (α) particles (He^{2+} ions), most of the particles penetrate the metal and suffer only small deflections from their original flight path. In 1909, at Rutherford's suggestion, H. Geiger and E. Marsden performed an experiment to see if any α particles were deflected at a large angle on striking a gold metal foil. A diagram of their experiment is shown in Figure 1–1. They discovered that some of the α particles actually were deflected by as much as 90° and a

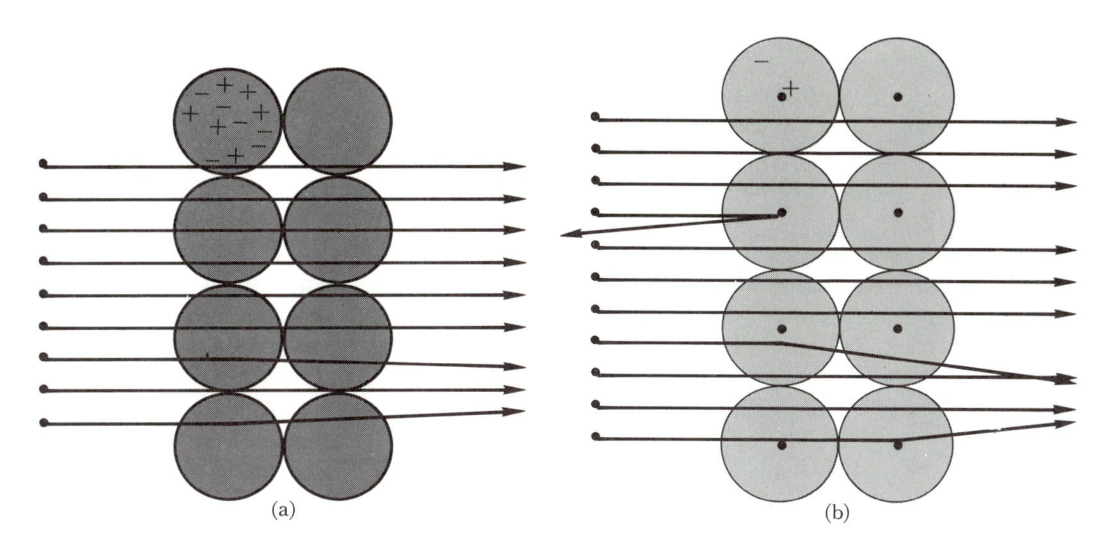

1–2
The expected outcome of the Rutherford scattering experiment, if one assumes (a) the Thomson model of the atom, and (b) the model deduced by Rutherford. In the Thomson model mass is spread throughout the atom, and the negative electrons are embedded uniformly in the positive mass. There would be little deflection of the beam of positively charged α particles. In the Rutherford model all of the positive charge and virtually all of the mass is concentrated in a very small nucleus. Most α particles would pass through undeflected. But close approach to a nucleus would produce a strong swerve in the path of the α particle, and head-on collision would lead to its rebound in the direction from which it came.

few by even larger angles. They concluded:

> If the high velocity and mass of the α particle be taken into account, it seems surprising that some of the α particles, as the experiment shows, can be turned within a layer of 6×10^{-5} cm of gold through an angle of 90°, and even more. To produce a similar effect by a magnetic field, the enormous field of 10^9 absolute units would be required.

Rutherford quickly provided an explanation for this startling experimental result. He suggested that atoms consist of a positively charged nucleus surrounded by a system of electrons, and that the atom is held together by electrostatic forces. The effective volume of the nucleus is extremely small compared with the effective volume of the atom, and almost all of the mass of the atom is concentrated in the nucleus. Rutherford reasoned that most of the α particles passed through the metal foil because the metal atom is mainly empty space, and that occasionally a particle passed close to the positively charged nucleus, thereby being severely deflected because of the strong coulombic (electrostatic) repulsive force. The Rutherford model of an atom is shown in Figure 1–2.

1–2 ATOMIC NUMBER AND ATOMIC MASS

Using the measured angles of deflection and the assumption that an α particle and a nucleus repel each other according to Coulomb's law of electrostatics, Rutherford was able to calculate the nuclear charge (Z) of the atoms of the elements used as foils. In a neutral atom the number of electrons equals the positive charge on the nucleus. This number is different for each element and is known as its *atomic number*.

The mass of an atomic nucleus is not determined entirely by the number of protons present. All nuclei, except for the lightest isotope of hydrogen, contain electrically neutral particles called neutrons. The mass of an isolated neutron is only slightly different from the mass of an isolated proton. The values of mass and charge for some important particles are given in Table 1–1.

Table 1–1. Some low-mass particles

Particle	Symbols	Mass in amu, atomic mass units[a]	Charge (esu × 10^{10})	Relative charge
Electron (β particle)	e^-, β^-, ${}_{-1}^{0}e$	0.0005486	−4.80325	−1
Proton	p, ${}_1^1p$, ${}_1^1H$	1.0072766	+4.80325	+1
Neutron	n, ${}_0^1n$	1.0086652	0	0
α Particle	α, ${}_2^4He$	4.00	+9.6065	+2
Deuteron	d, ${}_1^2d$, ${}_1^2p$	2.01	+4.80325	+1

[a] 1 amu = 1.660531×10^{-24} g.

1–3 NUCLEAR STRUCTURE

Rutherford's basic idea that an atom contains a dense, positively charged nucleus now is accepted universally. The extranuclear electrons on the average are far away from the nucleus, which accounts for an atom's relatively large effective diameter of about 10^{-8} cm. In contrast, the nucleus has a diameter of about 10^{-13} cm.

Recent scattering experiments have shown that the nuclear radius is roughly proportional to $Z^{1/3}$. These experiments have shown further that the arrangement of the nucleons (protons and neutrons) is structured and not random. Light nuclei are essentially all surface, whereas extremely heavy nuclei have a relatively thin outer layer. In the latter case there is evidence that the mutual electrostatic repulsion of the protons leads to a lower central density. It also has been postulated recently, from electron- and proton-scattering experiments, that essentially all nuclei have a neutron halo; that is, the neutron distribution in the nucleus extends beyond that of the protons.

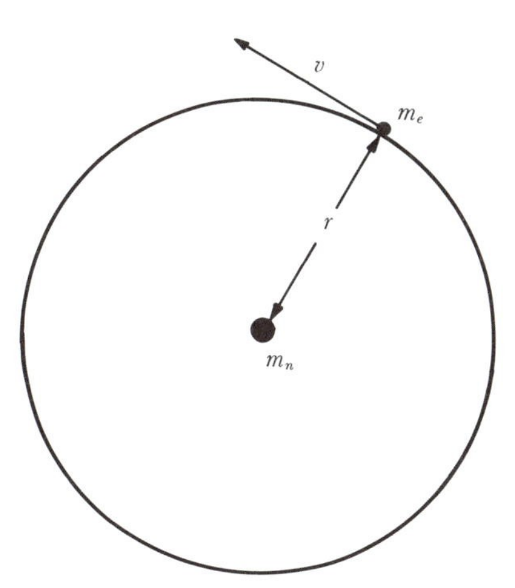

1–3
Bohr's picture of the hydrogen atom. A single electron of mass m_e moves in a circular orbit with velocity v at a distance r from a nucleus of mass m_n.

Additional information of interest has been derived from electron-scattering studies with very high-energy electrons. These studies indicate that the charge density (proton density) within the nucleus undergoes systematic fluctuations. These experiments are only the beginning of the investigation of the internal structure of atomic nuclei.

1–4 BOHR THEORY OF THE HYDROGEN ATOM

Immediately after Rutherford presented his nuclear model of the atom, scientists turned to the question of the possible structure of the satellite electrons. The first major advance was accomplished, in 1913, by a Danish theoretical physicist, Niels Bohr. Bohr pictured the electron in a hydrogen atom as moving in a circular orbit around the proton. Bohr's model is shown in Figure 1–3, in which m_e represents the mass of the electron, m_n the mass of the nucleus, r the radius of the circular orbit, and v the linear velocity of the electron.

For a stable orbit to exist the outward force exerted by the moving electron trying to escape its circular orbit must be opposed exactly by the forces of attraction between the electron and the nucleus. The outward force, F_0, is expressed as

$$F_0 = \frac{m_e v^2}{r} \tag{1–1}$$

This force is opposed exactly by the sum of the two attractive forces that keep the electron in orbit—the electrostatic force of attraction between the proton and the electron, plus the gravitational force of attraction. The electrostatic force is much stronger than the gravitational force, thus we may neglect the gravitational force. The electrostatic attractive force, F_e, between an electron of charge $-e$ and a proton of charge $+e$, is

$$F_e = -\frac{e^2}{r^2} \tag{1–2}$$

The condition for a stable orbit is that $F_0 + F_e$ equals zero:

$$\frac{m_e v^2}{r} - \frac{e^2}{r^2} = 0 \quad \text{or} \quad \frac{m_e v^2}{r} = \frac{e^2}{r^2} \tag{1–3}$$

Now we are able to calculate the energy of an electron moving in one of the Bohr orbits. The total energy, E, is the sum of the kinetic energy, KE, and the potential energy, PE:

$$E = KE + PE \tag{1–4}$$

in which KE is the energy due to motion,

$$KE = \frac{1}{2} m_e v^2 \tag{1–5}$$

and PE is the energy due to electrostatic attraction,

$$PE = -\frac{e^2}{r} \tag{1–6}$$

Thus the total energy is

$$E = \frac{1}{2} m_e v^2 - \frac{e^2}{r} \tag{1–7}$$

However, Equation 1–3 can be written $m_e v^2 = e^2/r$, and if e^2/r is substituted for $m_e v^2$ into Equation 1–7, we have

$$E = \frac{e^2}{2r} - \frac{e^2}{r} = -\frac{e^2}{2r} \tag{1–8}$$

Now we only need to specify the orbit radius, r, before we can calculate the electron's energy. According to Equation 1–8 atomic hydrogen should release energy continuously as r becomes smaller. To keep the electron from falling into the nucleus Bohr proposed a model in which the angular momentum of the orbiting electron could have only certain values. The result of this restriction is that only certain electron orbits are possible. According to Bohr's postulate the quantum unit of angular momentum is $h/2\pi$, in which h is the constant in Planck's famous equation $E = h\nu$. (E is energy in ergs, $h = 6.626196 \times 10^{-27}$ erg sec, and ν is frequency in $\sec^{-1}$.) In mathematical terms Bohr's assumption was that

$$m_e v r = n\frac{h}{2\pi} \tag{1–9}$$

in which $n = 1, 2, 3, \cdots$ (all integers to ∞). Solving for v in Equation 1–9 we can write

$$v = n\left(\frac{h}{2\pi}\right)\frac{1}{m_e r} \tag{1–10}$$

Substituting the value of v from Equation 1–10 into the condition for a stable orbit (Equation 1–3) we obtain

$$\frac{m_e n^2 h^2}{4\pi^2 m_e^2 r^2} = \frac{e^2}{r} \tag{1–11}$$

or

$$r = \frac{n^2 h^2}{4\pi^2 m_e e^2} \tag{1–12}$$

Equation 1–12 gives the radius of the possible electron orbits for the hydrogen atom in terms of the quantum number n. The energy associated with each possible orbit now can be calculated by substituting the value of r from Equation 1–12 into the energy expression (Equation 1–8), which gives

$$E_{\mathrm{H}} = \frac{2\pi^2 m_e e^4}{n^2 h^2} \tag{1–13}$$

Exercise. Calculate the radius of the first Bohr orbit.

Solution. The radius of the first orbit can be obtained directly from Equation 1–12,

$$r = \frac{n^2 h^2}{4\pi^2 m_e e^2}$$

Substituting $n = 1$ and the values of the constants we obtain

$$r = \frac{(1)^2(6.626196 \times 10^{-27}\ \text{erg sec})^2}{4(3.141593)^2(9.109558 \times 10^{-28}\ \text{g})(4.80325 \times 10^{-10}\ \text{esu})^2}$$

$$= 0.529177 \times 10^{-8}\ \text{cm} \cong 0.529\ \text{Å}$$

The Bohr radius for $n = 1$ is designated a_0.

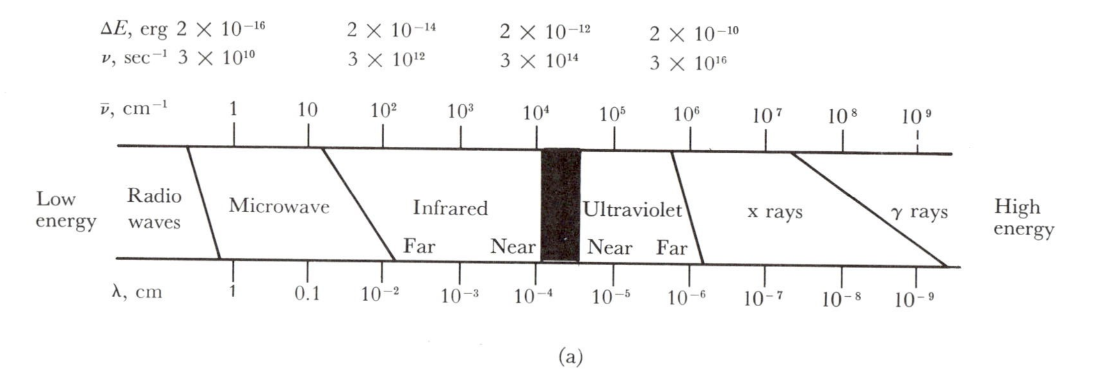

(a)

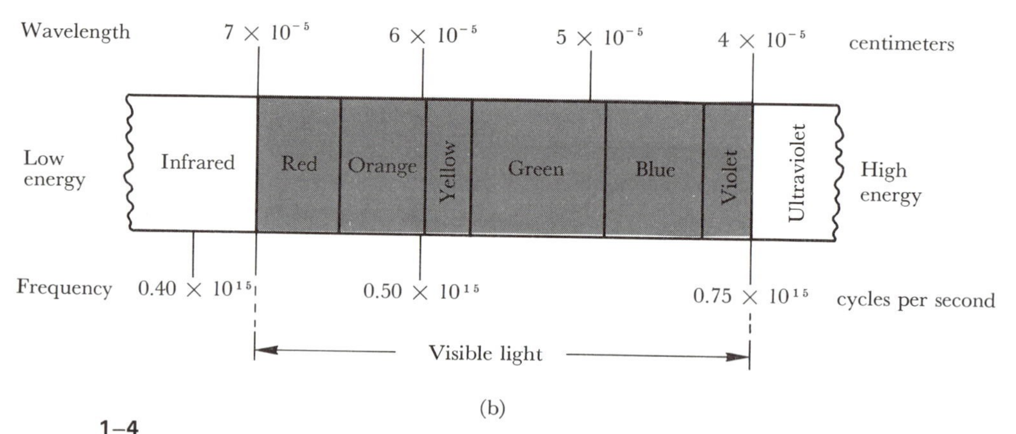

(b)

1–4
The spectrum of electromagnetic radiation. The visible region is only a small part of the entire spectrum. (a) Overall spectrum. (b) Visible region.

Exercise. Calculate the velocity of an electron in the first Bohr orbit of a hydrogen atom.

Solution. From Equation 1–10,

$$v = n\left(\frac{h}{2\pi}\right)\frac{1}{m_e r}$$

Substituting $n = 1$ and $r = a_0 = 0.529177 \times 10^{-8}$ cm we obtain

$$v = (1)(1.054592 \times 10^{-27}\ \text{erg sec}) \times \frac{1}{(9.109558 \times 10^{-28}\ \text{g})(0.529177 \times 10^{-8}\ \text{cm})}$$

$$= 2.187691 \times 10^{8}\ \text{cm sec}^{-1}$$

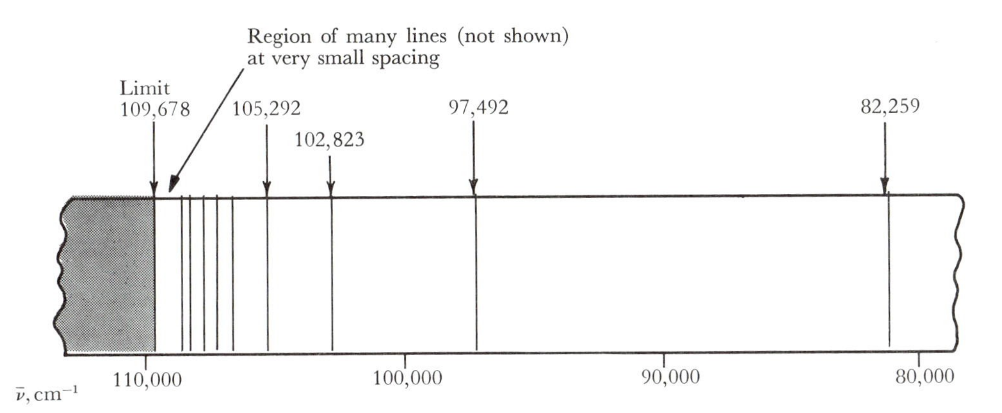

1–5
The electromagnetic absorption spectrum of hydrogen atoms. The lines in this spectrum represent ultraviolet radiation that is absorbed by hydrogen atoms as a mixture of all wavelengths is passed through a gas sample.

1–5 ABSORPTION AND EMISSION SPECTRA OF ATOMIC HYDROGEN

All atoms and molecules absorb light of certain characteristic frequencies. The pattern of absorption frequencies is called an *absorption spectrum* and is an identifying property of any particular atom or molecule. Because frequency is directly proportional to energy ($E = h\nu$), the absorption spectrum of an atom, such as hydrogen, shows that the electron can have only certain energy values, as Bohr proposed.

It is common practice to express positions of absorption in terms of the *wave number*, $\bar{\nu}$, which is the reciprocal of the wavelength, λ:

$$\bar{\nu} = \frac{1}{\lambda}$$

Because λ is related to frequency (ν) by the relationship $\lambda\nu = c$ (c = velocity of light = 2.9979×10^{10} cm sec^{-1}), we also have $\bar{\nu} = \nu/c$ and $E = h\bar{\nu}c$. Thus wave number and energy are directly proportional. If ν is in sec^{-1} units and c is in cm sec^{-1}, $\bar{\nu}$ is expressed in reciprocal centimeter (cm^{-1}) units. The spectrum of electromagnetic radiation is shown in Figure 1–4.

The absorption spectrum of hydrogen atoms is shown in Figure 1–5. The lowest-energy absorption corresponds to the line at 82,259 cm^{-1}. Notice that the absorption lines are crowded closer together as the limit of 109,678 cm^{-1} is approached. Above this limit absorption is continuous.

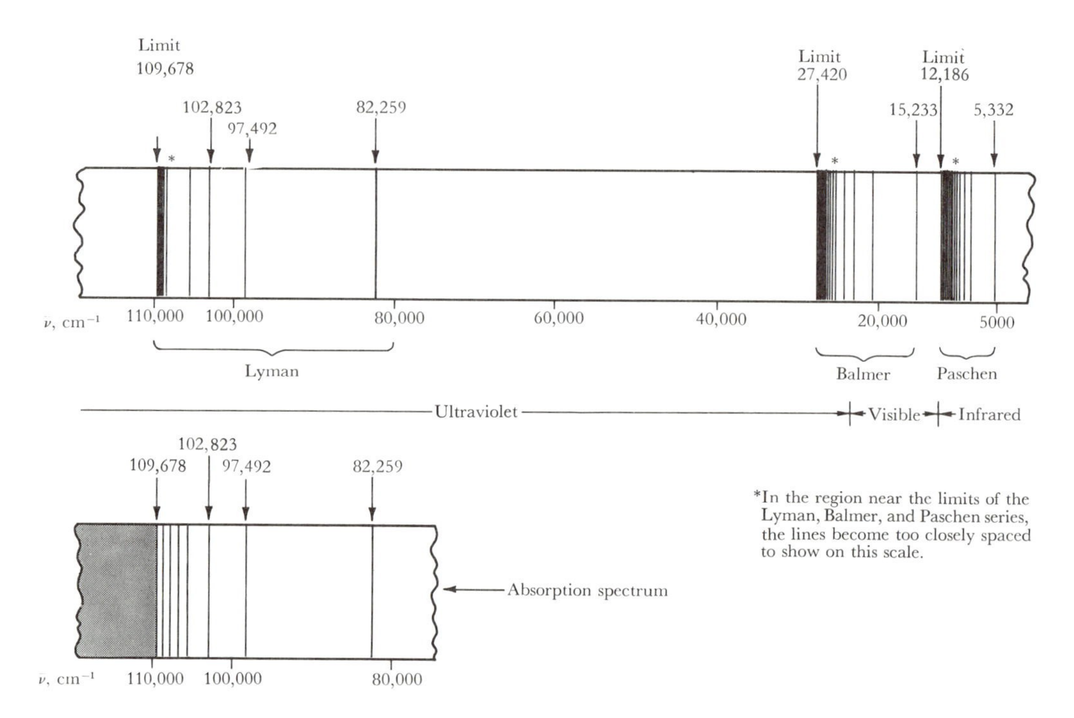

1–6
The emission spectrum from heated hydrogen atoms. The emission lines occur in series named for their discoverers: Lyman, Balmer, Paschen. The Brackett and Pfund series are farther to the right in the infrared region. The lines become more closely spaced to the left in each series until they finally merge at the series limit.

If atoms and molecules are heated to high temperatures, light of certain frequencies is emitted. For example, hydrogen atoms emit red light when heated. An atom that possesses excess energy (an "excited" atom) emits light in a pattern known as its *emission spectrum*. A portion of the emission spectrum of atomic hydrogen is shown in Figure 1–6. The emission spectrum contains more lines than the absorption spectrum. The lines in the emission spectrum at 82,259 cm^{-1} and above occur at the same positions as the lines in the absorption spectrum, but the emission lines below 82,259 cm^{-1} do not appear in the absorption spectrum (except at extremely high temperatures).

If we look more closely at the emission spectrum in Figure 1–6, we see that there are three distinct groups of lines. These three groups or series are named after the scientists who discovered them. The series that starts at 82,259 cm^{-1} and continues to 109,678 cm^{-1} is called the *Lyman series* and is in the ultraviolet portion of the spectrum. The series that starts at 15,233 cm^{-1} and continues to 27,420 cm^{-1} is called the *Balmer series* and covers a large portion of the visible and a small part of the ultraviolet spectrum. The lines between 5,332 cm^{-1} and 12,186 cm^{-1} are called the *Paschen series* and fall in the near-infrared spectral region.

Although the emission spectrum of hydrogen appears to be complicated, Johannes Rydberg formulated a fairly simple mathematical expression that gives all the line positions. This expression, called the *Rydberg equation*, is

$$\bar{\nu}_{\mathrm{H}} = R_{\mathrm{H}}\left(\frac{1}{n^2} - \frac{1}{m^2}\right)$$

In the Rydberg equation n and m are integers, with m greater than n; R_{H} is called the *Rydberg constant* and is known accurately from experiment to be 109,677.581 cm^{-1}.

Exercise. Calculate $\bar{\nu}_{\mathrm{H}}$ for the lines with $n = 1$ and $m = 2, 3,$ and 4.

Solution.

$n = 1, m = 2$ line:

$$\bar{\nu}_{\mathrm{II}} = 109{,}678\left(\frac{1}{1^2} - \frac{1}{2^2}\right) = 109{,}678\left(1 - \frac{1}{4}\right) = 82{,}259\ \mathrm{cm}^{-1}$$

$n = 1, m = 3$ line:

$$\bar{\nu}_{\mathrm{H}} = 109{,}678\left(\frac{1}{1^2} - \frac{1}{3^2}\right) = 109{,}678\left(1 - \frac{1}{9}\right) = 97{,}492\ \mathrm{cm}^{-1}$$

$n = 1, m = 4$ line:

$$\bar{\nu}_{\mathrm{H}} = 109{,}678\left(\frac{1}{1^2} - \frac{1}{4^2}\right) = 109{,}678\left(1 - \frac{1}{16}\right) = 102{,}823\ \mathrm{cm}^{-1}$$

We see that the preceding wave numbers correspond to the first three lines in the Lyman series. Thus we expect that the Lyman series corresponds to lines calculated with $n = 1$ and $m = 2, 3, 4, 5, \cdots$. Let us check this by calculating the wave number for the line with $n = 1, m = \infty$.

$n = 1, m = \infty$ line:

$$\bar{\nu} = 109{,}678(1 - 0) = 109{,}678\ \mathrm{cm}^{-1}$$

The wave number 109,678 cm^{-1} corresponds to the highest emission line in the Lyman series.

The wave number calculated for $n = 2$ and $m = 3$ is

$$\bar{\nu} = 109{,}678\left(\frac{1}{4} - \frac{1}{9}\right) = 15{,}233\ \mathrm{cm}^{-1}$$

This corresponds to the first line in the Balmer series. Thus the Balmer series corresponds to the $n = 2, m = 3, 4, 5, 6, \cdots$, lines. You probably would expect

1–7▶
Orbits and energy levels of the hydrogen atom. (a) Hydrogen-atom orbits. Each arc of a circle represents an electron radius for the electron outside the positive nucleus. Series for more energetic (excited) electrons dropping from outer levels to various inner levels are shown. (b) The energy changes for electrons dropping from excited energy states to less energetic levels. ΔE is determined by initial and final energies, which in turn are determined by the principal quantum number of the orbit, n.

the lines in the Paschen series to correspond to $n = 3, m = 4, 5, 6, 7, \cdots$; and they do. Now you should wonder where the lines are with $n = 4$, $m = 5, 6, 7, 8, \cdots$; and $n = 5$, $m = 6, 7, 8, 9, \cdots$. They are exactly where the Rydberg equation predicts they should be. The $n = 4$ series was discovered by Brackett and the $n = 5$ series was discovered by Pfund. The series with $n = 6$ and higher are located at very low frequencies and are not given special names.

The Bohr theory provided an explanation for the absorption and emission spectral lines in atomic hydrogen in terms of the orbits and energy levels shown in Figure 1–7. The orbit with $n = 1$ has the lowest energy, thus the one electron in atomic hydrogen occupies this level in its most stable state. The most stable electronic state of an atom or molecule is called the *ground state*. If the electron in a hydrogen atom can be only in certain orbits, it is easy to see why light is absorbed or emitted only at specific wave numbers. The absorption of light energy allows the electron to jump to a higher orbit. An excited hydrogen atom, in which the electron is not in the lowest-energy orbit, emits energy in the form of light when the electron falls back into a lower-energy orbit. The following series of spectral lines are a consequence of electronic transitions:

a) The Lyman series of lines arises from transitions from the $n = 2, 3, 4, \cdots$ levels into the $n = 1$ orbit.

b) The Balmer series of lines arises from transitions from the $n = 3, 4, 5, \cdots$ levels into the $n = 2$ orbit.

c) The Paschen series of lines arises from transitions from the $n = 4, 5, 6, \cdots$ levels into the $n = 3$ orbit.

The Bohr theory also explains why there are more emission lines than absorption lines. Excited hydrogen atoms may have $n = 2, 3, 4, 5, \cdots$. Any transition to a lower level is accompanied by emission of a quantum unit of light of energy $h\nu$. Such a light quantum is called a *photon*. Since transitions are possible to all orbits below the occupied orbit in the excited atom, an excited hydrogen atom with an electron in the $n = 6$ orbit, for example, may "decay" to a less excited state by the transitions $n = 6 \rightarrow n = 5$, $n = 6 \rightarrow n = 4$, $n = 6 \rightarrow n = 3$, and $n = 6 \rightarrow n = 2$, or it may decay to the ground state by the

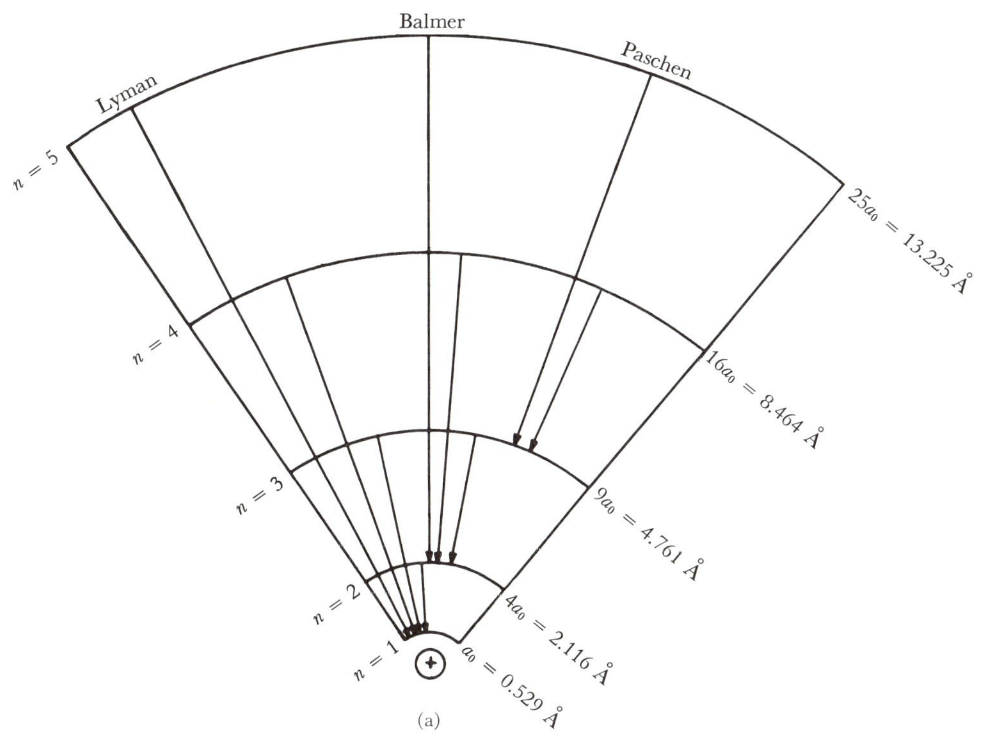
Balmer
Lyman
Paschen
n = 5
n = 4
n = 3
n = 2
n = 1
25a₀ = 13.225 Å
16a₀ = 8.464 Å
9a₀ = 4.761 Å
4a₀ = 2.116 Å
a₀ = 0.529 Å

(a)

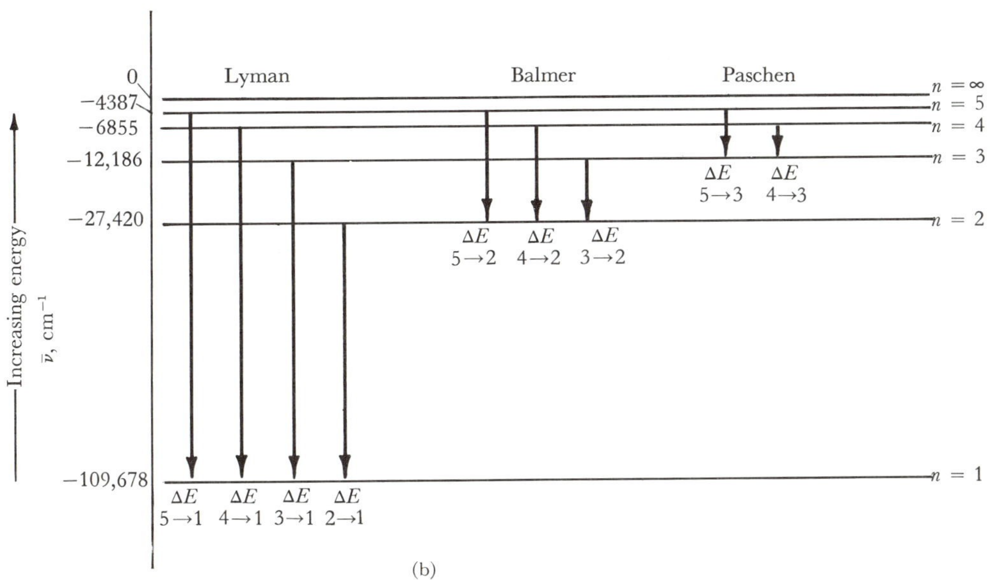
Increasing energy
ν̄, cm⁻¹
0
−4387
−6855
−12,186
−27,420
−109,678
Lyman
Balmer
Paschen
n = ∞
n = 5
n = 4
n = 3
n = 2
n = 1
ΔE 5→1
ΔE 4→1
ΔE 3→1
ΔE 2→1
ΔE 5→2
ΔE 4→2
ΔE 3→2
ΔE 5→3
ΔE 4→3

(b)

transition $n = 6 \rightarrow n = 1$. In other words, for the excited hydrogen atom with an $n = 6$ electron, there are five possible modes of decay, each with a finite probability of occurrence. Thus five emission lines, corresponding to five emitted photon energies ($h\nu$), result. The lines in the absorption spectrum of hydrogen are due to transitions from the ground state, $n = 1$ orbit, to higher orbits. Since almost no hydrogen atoms have n greater than one under typical absorption spectral conditions, only the transitions $n = 1 \rightarrow 2, 1 \rightarrow 3, 1 \rightarrow 4, 1 \rightarrow 5, \cdots$, and $1 \rightarrow \infty$ are observed. These are the absorption lines that correspond to the series of emission lines (the Lyman series) in which the excited states decay in one transition to the ground state.

Now we are able to derive the Rydberg equation from the Bohr theory. The transition energy (ΔE_H) of any electron jump in the hydrogen atom is the energy difference between an initial state, I, and a final state, II. That is,

$$\Delta E_H = E_{II} - E_I \qquad (1\text{–}14)$$

From Equation 1–13 the expression for the transition energy becomes

$$\Delta E_H = -\frac{2\pi^2 m_e e^4}{n_{II}^2 h^2} - \left(-\frac{2\pi^2 m_e e^4}{n_I^2 h^2}\right) \qquad (1\text{–}15)$$

or

$$\Delta E_H = \frac{2\pi^2 m_e e^4}{h^2}\left(\frac{1}{n_I^2} - \frac{1}{n_{II}^2}\right) \qquad (1\text{–}16)$$

Replacing ΔE_H with its equivalent wave number of light from the relationship $E = h\bar{\nu}c$ we can write

$$\bar{\nu}_H = \frac{2\pi^2 m_e e^4}{ch^3}\left(\frac{1}{n_I^2} - \frac{1}{n_{II}^2}\right) \qquad (1\text{–}17)$$

Equation 1–17 is equivalent to the Rydberg equation, with $n_I = n$, $n_{II} = m$, and $R_H = (2\pi^2 m_e e^4)/ch^3$. If we use the value of 9.109558×10^{-28} g for the rest mass of the electron, the Bohr-theory value of the Rydberg constant is

$$R_H = \frac{2\pi^2 m_e e^4}{ch^3} = \frac{2(3.141593)^2(9.109558 \times 10^{-28})(4.80325 \times 10^{-10})^4}{(2.997925 \times 10^{10})(6.626196 \times 10^{-27})^3}$$

$$= 109{,}737.3 \text{ cm}^{-1} \qquad (1\text{–}18)$$

Recall that the experimental value of R_H is 109,677.581 cm^{-1}. This remarkable agreement between theory and experiment was a great triumph for the Bohr theory.

1–6 IONIZATION ENERGY OF ATOMIC HYDROGEN

The *ionization energy* (IE) of an atom or molecule is the energy needed to remove an electron from the gaseous atom or molecule in its ground state, thereby forming a positive ion. For the hydrogen atom the process is

$$\mathrm{H}(g) \rightarrow \mathrm{H}^+(g) + e^-, \qquad \Delta E = IE$$

To calculate IE for hydrogen we can start with Equation 1–16,

$$\Delta E_{\mathrm{H}} = \frac{2\pi^2 m_e e^4}{h^2}\left(\frac{1}{n_{\mathrm{I}}^2} - \frac{1}{n_{\mathrm{II}}^2}\right)$$

For the ground state $n_{\mathrm{I}} = 1$, and for the state in which the electron is removed completely from the atom $n_{\mathrm{II}} = \infty$. Thus

$$IE = \frac{2\pi^2 m_e e^4}{h^2}$$

Recall that

$$a_0 = \frac{h^2}{4\pi^2 m_e e^2}$$

therefore

$$\frac{1}{2a_0} = \frac{2\pi^2 m_e e^2}{h^2}$$

Then substituting $1/2a_0$ into the expression for IE we have

$$IE = \frac{e^2}{2a_0} = \frac{(4.80325 \times 10^{-10}\ \mathrm{esu})^2}{2(0.529177 \times 10^{-8}\ \mathrm{cm})}$$

$$= 2.179914 \times 10^{-11}\ \mathrm{erg}$$

Ionization energies usually are expressed in electron volts (eV). Since 1 erg = 6.24145×10^{11} eV we calculate

$$IE = (2.179914 \times 10^{-11}\ \mathrm{erg})(6.24145 \times 10^{11}\ \mathrm{eV\ erg^{-1}}) = 13.60582\ \mathrm{eV}$$

The experimental value of the IE of the hydrogen atom is 13.598 eV.

1–7 GENERAL BOHR THEORY FOR A ONE-ELECTRON ATOM

The problem of one electron moving around any nucleus of charge $+Z$ is very similar to the hydrogen-atom problem. Since the attractive force is $-Ze^2/r^2$ the condition for a stable orbit is

$$\frac{m_e v^2}{r} = \frac{Ze^2}{r^2}$$

Proceeding from this condition in the same way as with the hydrogen atom we find

$$r = \frac{n^2h^2}{4\pi^2 m_e Z e^2}$$

and

$$E = -\frac{2\pi^2 m_e Z^2 e^4}{n^2 h^2}$$

Thus for the general case of nuclear charge $+Z$, for a transition from n_{I} to n_{II} we have

$$\Delta E = \frac{2\pi^2 m_e Z^2 e^4}{h^2}\left(\frac{1}{n_{\mathrm{I}}^2} - \frac{1}{n_{\mathrm{II}}^2}\right)$$

or simply

$$\Delta E = Z^2\,\Delta E_{\mathrm{H}}$$

Exercise. Calculate the third ionization energy of a lithium atom.

Solution. A lithium atom is composed of a nucleus of charge +3 ($Z = 3$) and three electrons. The *first ionization energy*, IE_1, of an atom with more than one electron is the energy required to remove one electron. For lithium

$$\mathrm{Li}(g) \rightarrow \mathrm{Li}^+(g) + e^-, \qquad \Delta E = IE_1$$

The energy needed to remove an electron from the unipositive ion, Li^+, is defined as the *second ionization energy*, IE_2, of lithium,

$$\mathrm{Li}^+(g) \rightarrow \mathrm{Li}^{2+}(g) + e^-, \qquad \Delta E = IE_2$$

and the *third ionization energy*, IE_3, of lithium is the energy required to remove the one remaining electron from Li^{2+}. For lithium, $Z = 3$ and $IE_3 = (3)^2(2.1799 \times 10^{-11}\ \mathrm{erg}) = 1.96191 \times 10^{-10}\ \mathrm{erg} = 122.45\ \mathrm{eV}$. The experimental value also is 122.45 eV.

The need for a better theory

The idea of an electron circling the nucleus in a well-defined orbit, analogous to the moon circling the earth, was easy to grasp, and Bohr's theory gained wide acceptance. However, it soon was realized that this simple theory was not the final answer. One difficulty was the fact that an atom in a magnetic field has a more complicated emission spectrum than the same atom in the absence of a magnetic field. This phenomenon is known as the *Zeeman effect* and it cannot be explained by the simple Bohr theory. However, the German physicist A. Sommerfeld was able to rescue temporarily the simple theory by suggesting elliptical orbits, in addition to circular orbits, for the electron. The combined Bohr–Sommerfeld theory explained the Zeeman effect very well.

A more serious problem was the inability of even the Bohr–Sommerfeld theory to account for the spectral details of the atoms that have several electrons. The theory also failed completely as a means of interpreting the periodic properties of the chemical elements. Thus the "orbit" approach soon was abandoned in favor of a powerful new theory of electronic motion based on the methods of wave mechanics.

1–8 MATTER WAVES

In 1924, the French physicist Louis de Broglie advanced the hypothesis that all matter possesses wave properties. He postulated that for every moving particle there is an associated wave with a wavelength given by the equation

$$\lambda = \frac{h}{mv} = \frac{h}{p} \tag{1–19}$$

The mass times the velocity, mv, is the momentum, p, of a particle and is a measure of the inertia or the tendency of a particle to remain in motion.

In 1927, C. Davisson and L. H. Germer demonstrated that electrons are diffracted by crystals in a manner similar to the diffraction of x rays. These electron-diffraction experiments dramatically supported de Broglie's postulate of matter waves. Thus modern experimental evidence indicates that "particle" and "wave" phenomena are not mutually exclusive, rather they emphasize different attributes of all matter. Therefore it is reasonable to expect that all things in nature possess both the properties of particles (discrete units), and the properties of waves (continuity). The particle aspect of matter is more important in describing the properties of the relatively large objects encountered in normal observations. The wave properties are more important in describing the characteristics of many of the extremely minute objects outside the realm of ordinary perception.

It is helpful to point out why the characteristics of wave motion are not apparent in the motion of easily observable objects. Everything from a baseball to a battleship has a wavelike nature associated with its movements. However, for relatively large objects the wavelengths are so small that we cannot perceive them. For example, consider a baseball with a mass of 200 g and a speed of 3.0×10^3 cm sec^{-1}. From de Broglie's equation the wavelength of the associated wave is $6.63 \times 10^{-27}/(200)(3.0 \times 10^3)$ or approximately 10^{-32} cm. It is apparent that this wavelength is too small for ordinary observation. In contrast, an electron with a rest mass of approximately 10^{-27} g moving at the same speed would have a wavelength of about 2×10^{-3} cm, which is well within the realm of ordinary observation.

De Broglie's hypothesis of matter waves provided the foundation for an entirely new theory, which firmly established the quantum properties of energy in physical systems. The new theory logically is called *quantum mechanics.* The remainder of this book will be devoted to the application of quantum mechanics to the description of electronic structures of atoms and molecules.

1–9 THE UNCERTAINTY PRINCIPLE

One of the most important consequences of the dual nature of matter is the uncertainty principle, which was derived in 1927 by Werner Heisenberg. The *Heisenberg uncertainty principle* states that it is impossible to know simultaneously both the momentum and the position of a particle with certainty. This means that as a measurement of momentum or velocity is made more precisely, a measurement of the position of the particle is correspondingly less precise. Similarly, if the position is known precisely, the momentum must be less well known. Heisenberg showed that the lower limit of this uncertainty is Planck's constant divided by 4π. This relationship is expressed as

$$(\Delta p_x)(\Delta X) \geqq \frac{h}{4\pi} \tag{1–20}$$

in which Δp_x is the uncertainty in the momentum along the X direction and ΔX is the uncertainty in the position. Equation 1–20 is only one of several ways of expressing the uncertainty principle. For example, we could write

$$(\Delta p_y)(\Delta Y) \geqq \frac{h}{4\pi} \qquad \text{and} \qquad (\Delta p_z)(\Delta Z) \geqq \frac{h}{4\pi}$$

The uncertainty principle can be clarified by considering an attempt to measure both the position and the momentum of an electron (Figure 1–8). In this attempt the position of the electron is to be pinpointed by using electromagnetic radiation as the measuring device. Thus a photon must collide with and be reflected from the electron for that electron to be "seen." But since photons and electrons have nearly the same energy, collisions between them result in changes in the electron's velocity and consequently in its momentum.

The position of a particle at any instant can be observed with a precision comparable to that of the wavelength of the incident light. However, during that measurement the light photon interacting with the electron causes an alteration of the electron's momentum of the same magnitude as the momentum of the photon itself. To increase the precision in detecting the position of the electron, the wavelength of the incident light must be decreased. But as the wavelength decreases the frequency of the light and the energy of the corresponding photons increase, thereby increasing the uncertainty in the momentum of the electron. Therefore uncertainty is inherent in the very nature of the measuring process.

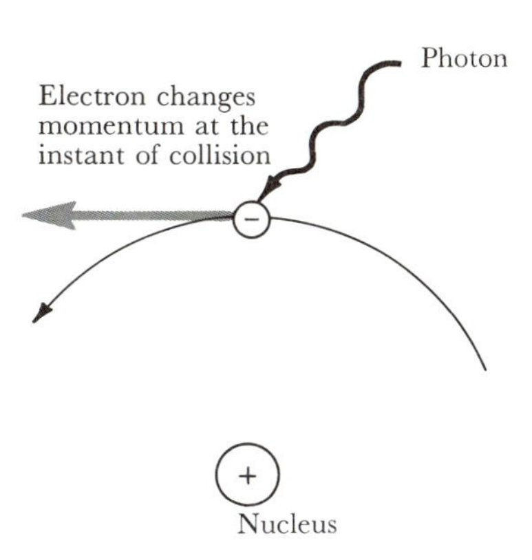

1–8
The position of an electron at an instant of time should be determinable by a "super microscope" with light of small wavelength, λ (x rays or γ rays). However, light photons of small λ have great energies and therefore very large momenta. A collision of one of these photons with an electron instantly changes the electronic momentum. Thus as the position is better resolved, the momentum becomes more and more uncertain.

Consider an attempt to measure simultaneously the momentum and position of an electron. For example, assume that the electron is moving with a velocity of 10^7 cm sec^{-1}. Suppose that an attempt to measure its position is made with visible light of frequency 10^{15} sec^{-1}. The energy of the electron is simply its kinetic energy or energy of motion, which is $1/2mv^2$ or approximately 10^{-12} erg. The energy of light photons is given by $E = h\nu$. From this equation, with visible light of frequency 10^{15} sec^{-1} the energy of a photon also is about 10^{-12} erg. Therefore, in a measurement collision the momentum of the electron will be changed instantaneously, with a corresponding large uncertainty in the momentum at that instant.

The uncertainty principle need not be considered in measurements of the momentum and position of large bodies, because the measurements are not sufficiently precise to observe any inherent uncertainty. Only for small particles does the uncertainty principle become important. The following comparison of a baseball and an electron illustrates this.

Recall the baseball weighing 200 g and traveling with a velocity of 3.0×10^3 cm sec^{-1} (Section 1–8). This baseball has a momentum of 6.0×10^5 g cm sec^{-1}. Furthermore, consider attempts to measure both the position and the momentum of the baseball. Suppose that it is possible to measure the momentum with a precision of one part in a trillion (10^{12}). This would mean that the momentum is known with an uncertainty no greater than 6.0×10^{-7} g cm sec^{-1}. The uncertainty principle asserts that the position then can be known with a precision of approximately 10^{-21} cm, which generally is much greater precision than typical physical measurements can achieve.

Now consider the electron with a mass of 10^{-27} g moving with the same velocity as the baseball. The electron would have a momentum of $3.0 \times$

10^{-24} g cm sec^{-1}. If it were possible to measure that momentum with the same precision as in the case of the baseball, Δp would be 3.0×10^{-36} g cm sec^{-1} and the uncertainty in position, ΔX, would be close to a billion centimeters. From these examples it is clear that the uncertainty principle is important only in considering measurements of the small particles that comprise an atomic system.

1–10 ATOMIC ORBITALS

If electrons moved in simple orbits, then the momentum and position of the electron could be determined exactly at any instant. According to the uncertainty principle this situation does not correspond to reality. Therefore in discussing the motion of an electron of known energy or momentum around a nucleus, it is necessary to speak only in terms of the probability of finding that electron at any particular position.

To be consistent with the idea of the dual nature of matter we consider the motion of an electron as a wave. According to de Broglie's equation the velocity (consequently the energy) of the electron determines the wavelength and therefore the frequency of the associated wave. The amplitude of the wave in any region indicates the relative probability of finding an electron in that region.

To clarify the concept of electron probability it is helpful to do a hypothetical experiment in which we take a set of instantaneous pictures of an electron with a specific energy moving around a nucleus. If all these imaginary sequential pictures, with the electron appearing as a small dot in each picture, were superimposed, a "cloud" similar to that shown in Figure 1–9(a) would result. This picture is called an electron-density or electron-cloud representation. The density of dots in a given spatial region is a pictorial representation of the *probability density* in that region.

If the electron cloud in Figure 1–9(a) is examined, the probability density is seen to be greater near the nucleus and to decrease as the distance from the nucleus increases. Consequently an atom cannot be given a definite radius, rather it is characterized by a fuzzy electron cloud possessing no definite boundaries. However, electronic motion with a poorly defined boundary is unwieldy for efficient pictorial representation. Thus, as an arbitrary boundary for the electronic motion, it is convenient to set a volume with a surface along which the probability density is some constant value. This is called a *boundary surface of constant probability density*. This ensures that the shape of electron density will be represented accurately. In addition, it is common to define this surface as enclosing about 90–99% of the electron density.

For the lowest-energy orbital of the hydrogen atom the boundary surface is a sphere of approximately 10^{-8}-cm diameter. In Figure 1–9(b) a circular cross section of the boundary surface is superimposed over the electron-density

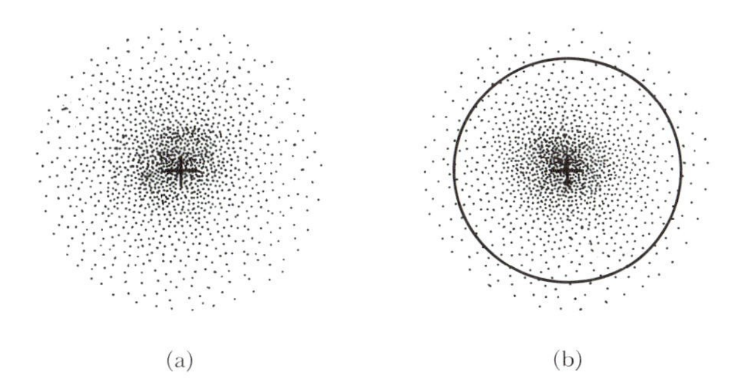

1–9
(a) Electron-density or electron-cloud representation of the motion of an electron around a positive nucleus. (b) Circular cross section of a spherical boundary surface enclosing 90–99% probability region of a 1*s* orbital. If an experimentalist could make a large number of measurements of the position of the electron, 90–99% of the time the electron would be found within the sphere enclosed by the boundary surface.

picture. In place of the model using planetary orbits, the probability-density representation describes the motion of the electron as being concentrated in a certain region with a certain shape around the nucleus. We will associate the name "electron orbital" with this type of model of electron motion. An orbital can be represented by a mathematical function, as well as pictorially by one of the probability-density representations that we have discussed. The mathematical functions that represent electron orbitals and their relationship to probability densities are presented in the next section.

1–11 THE WAVE EQUATION AND QUANTUM NUMBERS FOR THE HYDROGEN ATOM

In 1926, Erwin Schrödinger advanced the famous wave equation that relates the energy of a system to its wave properties. Because of the rather complicated mathematical form of the Schrödinger equation and its solutions, we shall discuss the important results only qualitatively.

The Schrödinger equation for the system of one proton and one electron (i.e., the hydrogen atom) can be solved exactly. A quantization results in which only certain orbitals and energies are possible. These orbitals are specified by three orbital quantum numbers. The first orbital quantum number is called the *principal quantum number* and is given the symbol n. In a general way, the effective volume of an electron orbital depends on n. The orbital energy is determined by the value of n according to the expression

$$E = -\frac{k}{n^2}$$

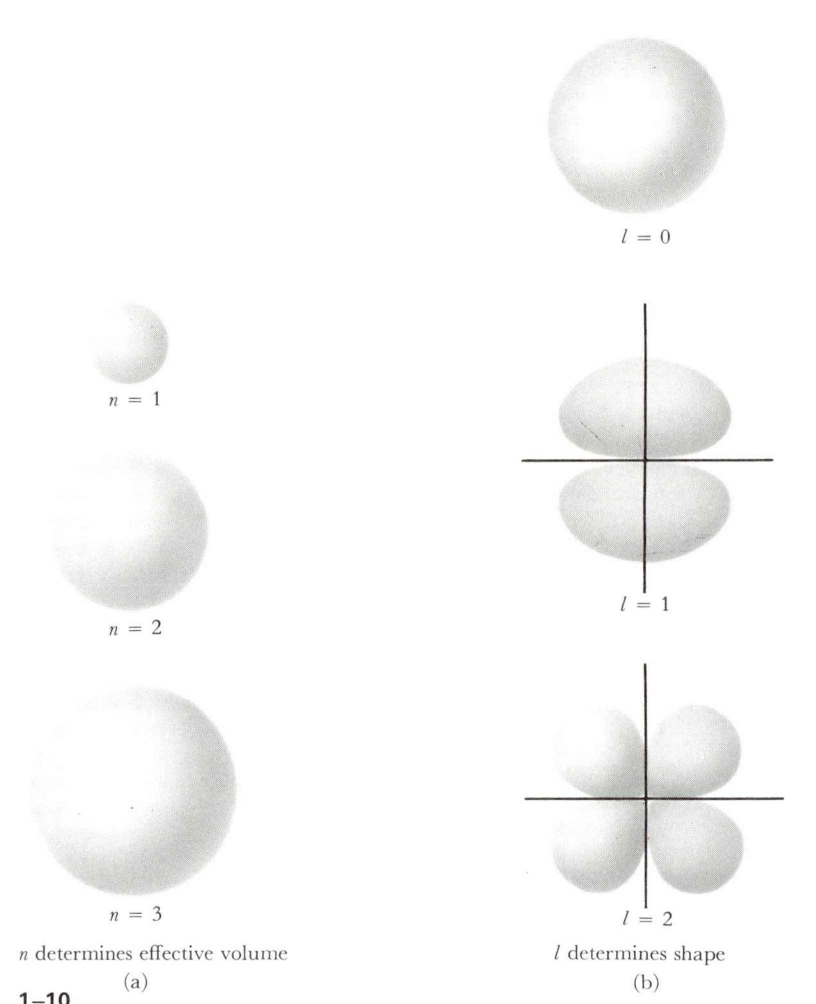

1–10
Summary of the most important aspects of the hydrogen orbitals. (a) The principal quantum number, n, indicates approximately the effective volume of the orbital. (b) The orbital-shape quantum number, l, determines the general shape of the orbital.

in which $k = 2\pi m_e e^4/h^2$, as in the Bohr theory. A very useful value for k is 13.6 eV.

The second orbital quantum number is designated l. The l quantum number determines, in a general way, the shape of the region in which the electron moves. Therefore l is called the *orbital-shape quantum number*. These first two quantum numbers together determine the spatial properties of the electron orbital, as shown in Figure 1–10.

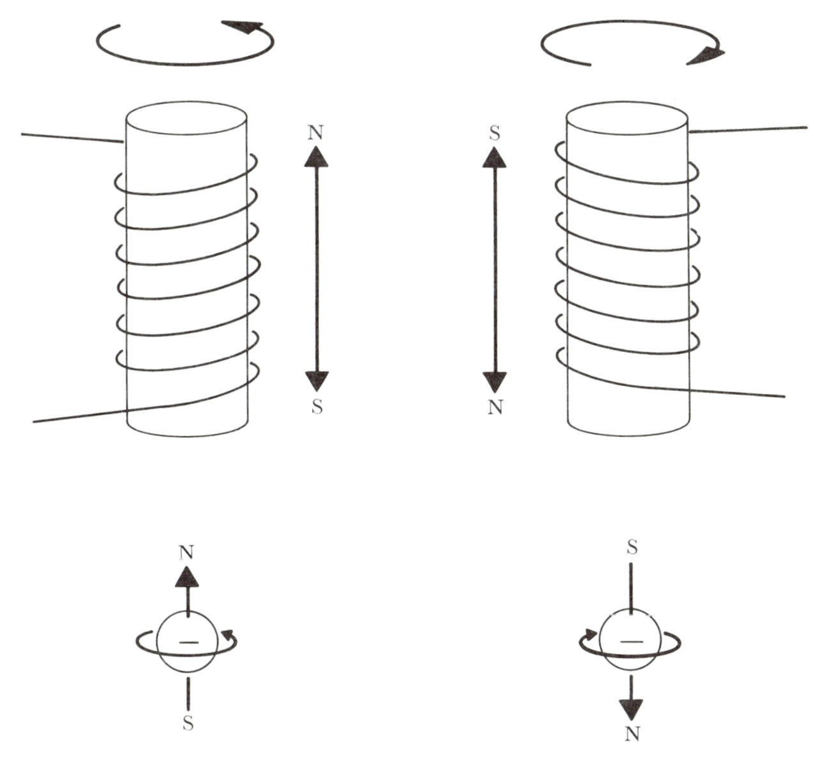

1–11
Electron spin in a magnetic field. Just as the direction of current flow around an iron bar determines the direction of the polarity of the magnet induced in the iron bar, so the direction of the spin of an electron determines its spin quantum number. Electron spin is quantized in half-integer units and for each electron can have values of $\pm 1/2$.

The third orbital quantum number, m_l, determines the orientation of a particular spatial configuration in relation to an arbitrary direction. The introduction of an external magnetic field most conveniently provides the arbitrary reference axis. We call the third quantum number the *orbital-orientation quantum number*. The preceding three orbital quantum numbers determine explicitly an electron orbital.

An electron moving in a particular orbital has properties that can be explained by imagining that the electron is a tiny bar magnet with a north and a south pole. This "bar-magnet" behavior is described as *electron spin* and can be in either of two directions in reference to an arbitrary axis. The *spin quantum number*, m_s, specifies the direction of spin in space with reference to such an arbitrary axis. The two orientations of electron spin in a magnetic field are shown schematically in Figure 1–11.

Solution of the Schrödinger equation imposes certain restrictions on the values that the four quantum numbers can have. The principal quantum number can have any positive integral value from one to infinity. But a particular value of the principal quantum number determines the possible values of the l quantum number. Correspondingly, each value of the l quantum number determines the possible values of the m_l quantum number. The four quantum numbers and the values that each can have are:

n = principal quantum number = $1, 2, 3, \cdots, \infty$

l = orbital-shape quantum number = $0, 1, 2, \cdots, n - 1$ (for any value of n)

m_l = orbital-orientation quantum number = $-l, -l + 1, \cdots, 0, \cdots, l - 1, l$ (for any value of l)

m_s = spin quantum number = $\pm\frac{1}{2}$

Exercise. An electron has the principal quantum number four. What are the possible values of l, m_l, and m_s for this electron?

Solution. With $n = 4$, l may have a value of 3, 2, 1, or 0.

For $l = 3$ there are seven possible values for m_l: $3, 2, 1, 0, -1, -2, -3$.

For $l = 2$ there are five possible values for m_l: $2, 1, 0, -1, -2$.

For $l = 1$ there are three possible values for m_l: $1, 0, -1$.

For $l = 0$ there is only one possible value of m_l: 0.

Since for each set of orbital quantum numbers the electron can have either $+1/2$ or $-1/2$ spin, there are 32 possible combinations of l, m_l, and m_s with $n = 4$.

The Schrödinger wave equation can be set up for any atom, but it can be solved exactly only for the hydrogen atom (or ions with one electron). Consequently the four quantum numbers, and the corresponding wave functions obtained by solving the Schrödinger equation of motion, actually apply only to "hydrogenlike" atoms. Nevertheless, the concept of quantum numbers and orbitals remains very useful. To describe many-electron atoms we assume orbitals that are analogous to hydrogenlike orbitals, but modify them somewhat because of the repulsive interactions of the electrons.

Quantum number specification of orbitals

The different n values for the hydrogen atom are called *energy levels* or "*shells*." For each different shell (or n value) the l quantum number can have only certain discrete values, corresponding to orbitals of different shapes. For historical reasons the 0, 1, 2, and 3 values of the l quantum number are designated by the letters *s*, *p*, *d*, and *f*, respectively. The combination of the principal quantum number and the appropriate letter corresponding to the value of the l quantum number is the shorthand notation for a particular orbital. For example, the combination $n = 1$, $l = 0$ is a 1*s* orbital, and that of $n = 3$, $l = 1$ is a 3*p* orbital.

The *wave function* ψ, a particular solution of the Schrödinger wave equation, mathematically describes the motion of an electron in an orbital. The amplitude of the wave function, indicated by its magnitude at various points in the region around a nucleus, gives the approximate probability of finding an electron in that orbital at any particular point. However, the precise value of the wave function squared (ψ^2) is a direct measure of the probability density of the electron at any point in space. In other words, the greater the amplitude or value of ψ in a given region, the greater the probability that the electron is in that region.

For the lowest-energy level or shell, n is one. The rules relating the values of the quantum numbers require that the l and m_l quantum numbers be zero. Accordingly, there is only one orbital with $n = 1$ ($l = 0$, $m_l = 0$). This orbital is designated as 1*s* and is described by the following wave function, ψ:

$$\psi(1s) = Ne^{-r} \tag{1–21}$$

in which

$$N = \frac{1}{\pi^{1/2}}$$

In Equation 1–21 the symbol $\psi(1s)$ designates the wave function of the electron in the $n = 1$ orbital. The normalization constant, N, is fixed so that the probability of finding the electron somewhere in space is one. The quantity e is the base of natural or Naperian logarithms and is approximately 2.72. The distance r from the nucleus is expressed in atomic units, that is, in units of a_0, the Bohr radius. The value of the wave function as a function of the radial distance from the center of the atom, r, is shown in Figure 1–12. The $\psi(r)$ function is for the hydrogen 1*s* electron orbital. The electron density associated with an electron orbital is obtained from the square of the wave function, $\psi^2(r)$, which gives the probability density, $P(r)$, for the electron at a given point in space:

$$P(r) = N^2e^{-2r} = \frac{1}{\pi}e^{-2r} \tag{1–22}$$

In Figure 1–13 the probability density for a 1*s* orbital is plotted as a function of r. For *s* orbitals the probability of finding an electron at the nucleus is finite (nonzero), whereas for all other orbitals the value of $\psi^2(r)$ at the nucleus is zero.

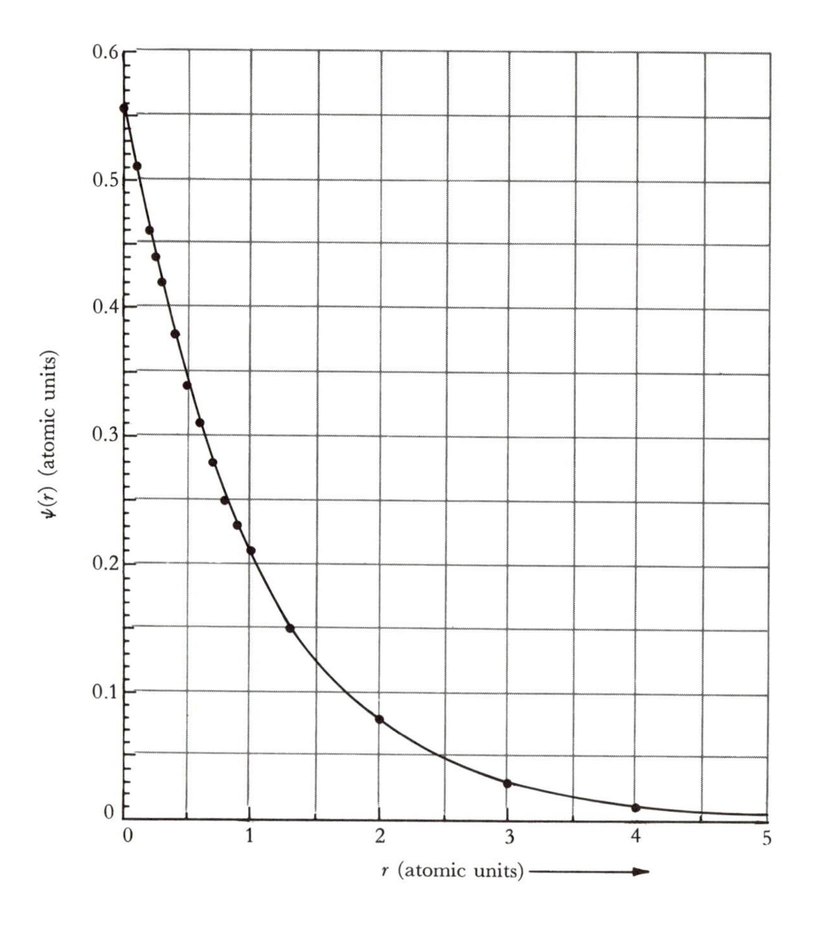

1–12
Plot of $\psi(1s) = Ne^{-r}$ for atomic hydrogen. The magnitude of the wave function, $\psi(r)$, gives approximately the chance of finding the 1*s* electron at any distance *r* from the nucleus. The distance *r* is measured in atomic units, that is, in units of a_0, the Bohr radius ($1a_0 = 0.529$ Å).

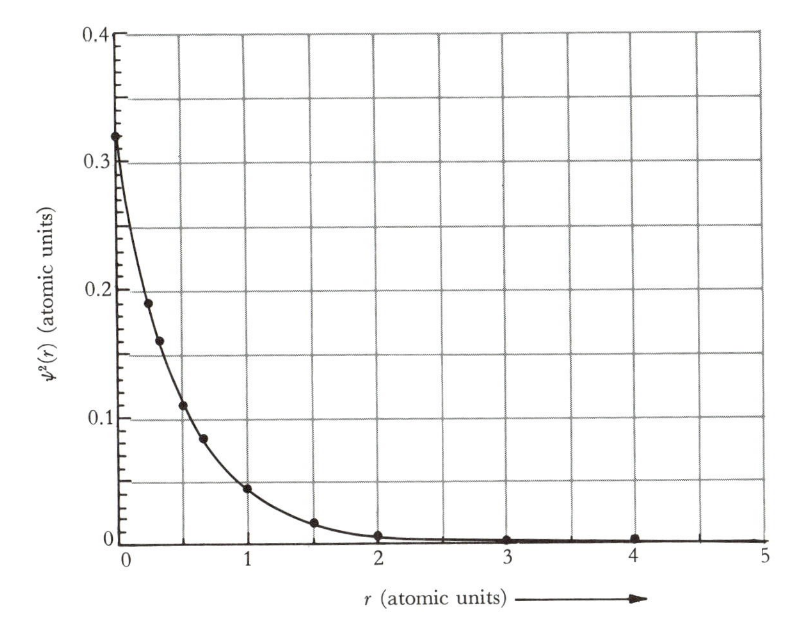

1–13
Plot of $\psi^2(r) = P(r) = N^2e^{-2r}$ for atomic hydrogen. The precise values of the square of the wave function is a direct measure of the probability density of an electron at any distance r from the nucleus. The probability curve never reaches zero, even at $r = \infty$. However, the sphere around the nucleus that contains 99% of the probability [see Figure 1–9(b)] has a radius of 4.2 atomic units (2.2 Å).

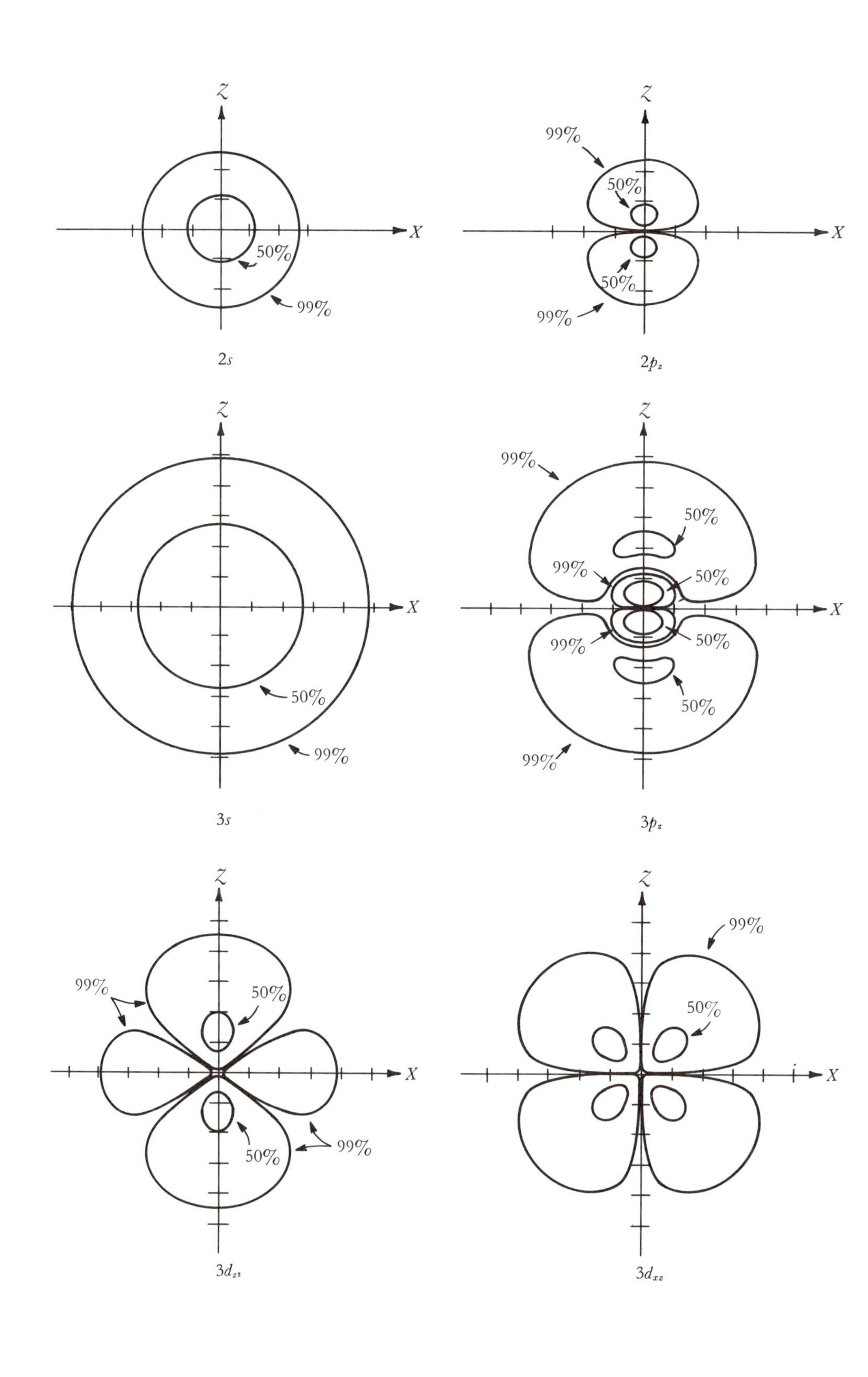

2s

$2p_z$

3s

$3p_z$

$3d_{z^2}$

$3d_{xz}$

◀1–14
Contour diagrams in the XZ plane for hydrogen wave functions that show the 50% and 99% contours. X and Z axes are marked in intervals of five atomic units. The $3p_z$ orbital differs from the $2p_z$ in having another nodal surface as a spherical shell around the nucleus at a distance of approximately six atomic units.

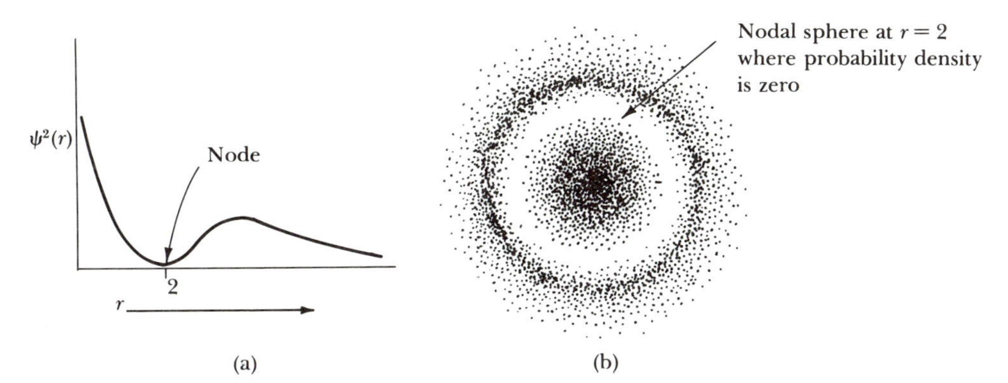

1–15
The hydrogen 2s orbital. (a) The graph of $\psi^2(r)$ against r. (b) A cross section through the probability function plotted in three dimensions. Probability density is represented by stippling.

The electron density of a 1s orbital is illustrated in Figure 1–9(a). The density of dots is a pictorial representation of the probability density. To restrict the region of space referred to in the discussion of orbitals, a cross-sectional contour of constant probability density can be used and is shown in Figure 1–9(b) for the 1s orbital. In this case the contour is a circle, which represents a cross section of the spherical 1s orbital. The contour is drawn such that the electron is inside the sphere for which the circle is a cross section about nine tenths of the time (or the charge inside the sphere is $-0.9e$). Figure 1–14 shows two contours for each of several other hydrogen atomic orbitals.

When the principal quantum number is 2 the l quantum number is restricted to two values, 0 and 1. The orbital with $l = 0$ is the 2s orbital and can be described analytically as

$$\psi(2s) = \frac{1}{4(2\pi)^{1/2}}(2 - r)e^{-r/2} \qquad (1\text{–}23)$$

The important differences between the 2s and 1s orbitals for the hydrogen atom are that the 2s orbital is effectively larger than the 1s orbital, and that for $r = 2$ the 2s wave function is zero. A surface on which a wave function is zero is called a *node* (or nodal surface). Thus the 2s orbital of atomic hydrogen has a nodal sphere with a radius of two atomic units, as shown in Figure 1–15.

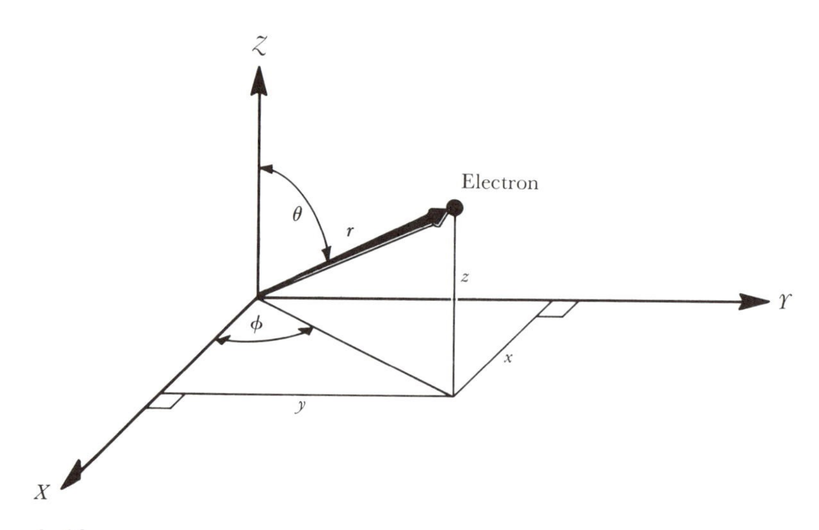

1–16
The polar coordinates θ, ϕ, and r and their relationship to the X, Y, and Z axes. It can be shown that $x = r \sin\theta \cos\phi$, $y = r \sin\theta \sin\phi$, and $z = r\cos\theta$.

It is important to remember that all s orbitals are spherically symmetrical. Orbitals with $l = 0$ and different values of n differ only in "effective volume," and in the number of nodes.

In the second shell, with $n = 2$, an orbital with $l = 1$ is encountered. There are three orbitals with $l = 1$ because the m_l quantum number can have the values $-1, 0,$ and 1. Unlike the s orbitals, which are spherically symmetrical, the three $2p$ orbitals have directional properties. Accordingly, the $2p$ orbital with $m_l = 0$ is specified in the polar coordinates of Figure 1–16 as

$$\psi(2p_z) = \frac{1}{4(2\pi)^{1/2}}(\cos\theta)re^{-r/2} \quad (1\text{–}24)$$

The $2p_z$ orbital, which is described by Equation 1–24, has regions of greatest concentration or probability along the Z axis. The electron-density representation of this $2p$ orbital is shown in Figure 1–17(a). Examining this representation we see that the probability of finding the electron in the XY plane is zero [i.e., $\psi(2p_z) = 0$ when $\cos\theta = 90°$]. This nodal plane containing the atomic nucleus is a property of all p orbitals.

The $2p_z$ orbital also can be described with a contour diagram or a spatial representation. We can find the contour diagram by plotting lines of constant probability density, which correspond to constant $|\psi|$ or ψ^2. The square of the $\psi(2p_z)$ function is given by the equation

$$P(r) = [\psi(2p_z)]^2 = \frac{r^2 \cos^2\theta e^{-r}}{32\pi} \quad (1\text{–}25)$$

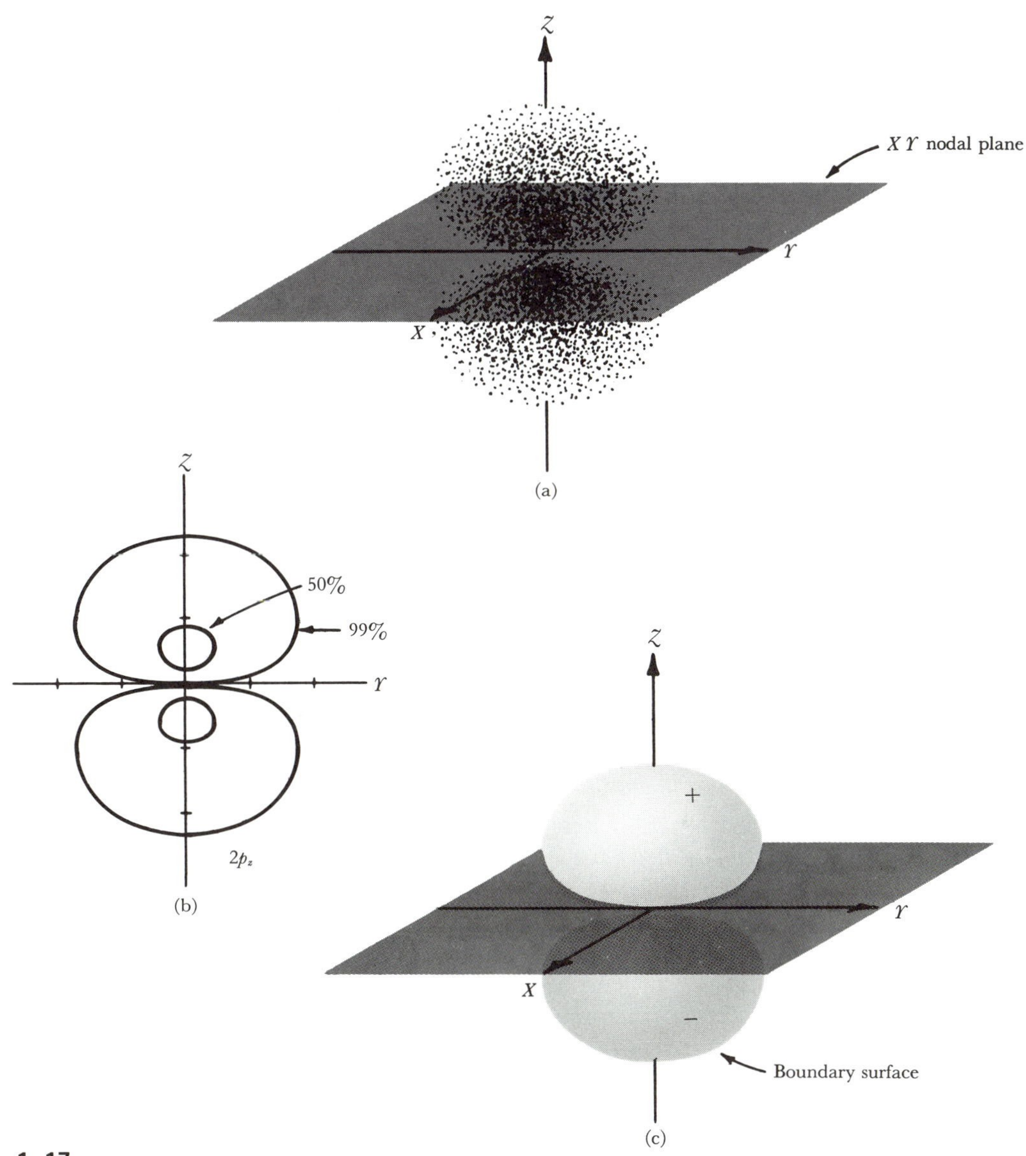

1–17
Three ways of representing the $2p_z$ atomic orbital of hydrogen. (a) ψ^2 represented by stippling. (b) Contour diagram of the $2p_z$ orbital. The contours represent lines of constant ψ^2 in the *YZ* plane and have been chosen so that, in three dimensions, they enclose 50% or 99% of the total probability density. The $2p_z$ orbital is symmetrical around the *Z* axis. (c) The 99% probability shell portrayed as a surface. The plus and minus signs on the two lobes represent the relative signs of ψ and should not be confused with electric charge. Notice that there is no probability of finding the electron on the *XY* plane. Such a surface, which need not be planar, is called a nodal surface.

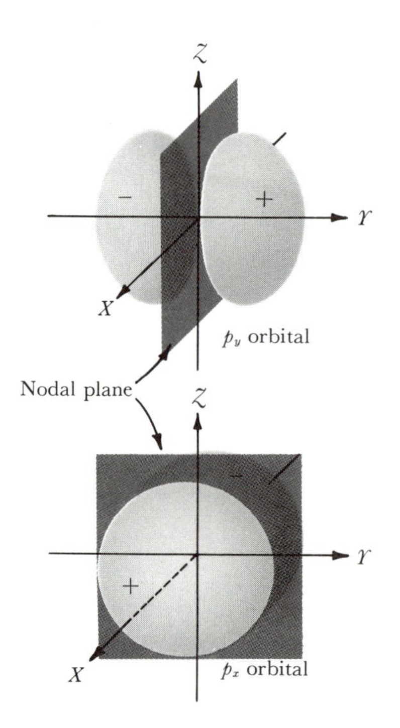

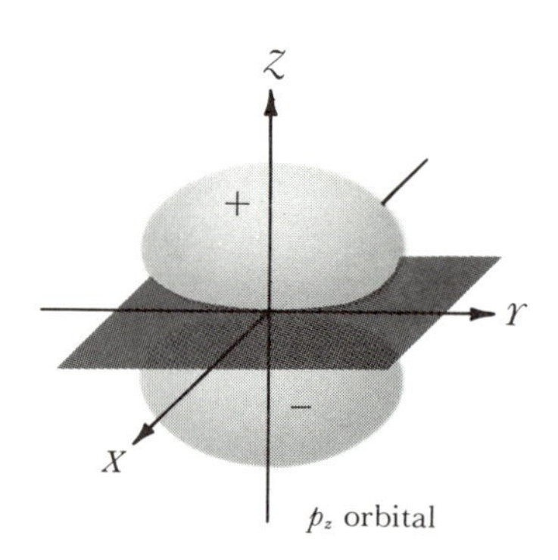

1–18
Boundary surfaces enclosing 99% of the probability for the $2p_y$, $2p_x$, and $2p_z$ orbitals of hydrogen. Notice the nodal plane of zero probability density in each orbital.

The resulting contours and a spatial representation for the $2p_z$ orbital are shown in Figure 1–17(b, c).

The signs of the two lobes in the spatial representation are a reminder that the $2p_z$ wave function is positive for positive Z values and negative for negative Z values; that is, positive for θ values between 0° and 90° and negative for θ values between 90° and 180°. All p functions change sign when inverted at the atomic nucleus and are said to be antisymmetric. In contrast, s orbital functions are symmetrical because inversion does not generate a change in algebraic sign. The symmetry properties of orbitals are emphasized here because, as we will see later, they are important in classifying bonds between atoms.

A complete set of p orbitals is shown in Figure 1–18. The three equivalent p orbitals differ only in their spatial orientations. They are designated p_x, p_y, and p_z, depending on their directional properties with respect to the X, Y, and Z axes.

When the principal quantum number is three, l can have three values: 0, 1, and 2. A $3s$ orbital has $l = 0$ and is similar to the s orbitals described previously. (It is a general result that the number of nodes equals $n - 1$; thus the $3s$ orbital has two nodes.) When $l = 1$ there are three $3p$ orbitals. The $2p$ and $3p$ orbitals with the same m_l quantum number have the same angular dependence, although the boundary contours of the $3p$ orbitals are more complicated than those for the $2p$ orbitals because of the presence of an additional node. However, the outer part of a $3p$ orbital looks like a $2p$ orbital (see Figure

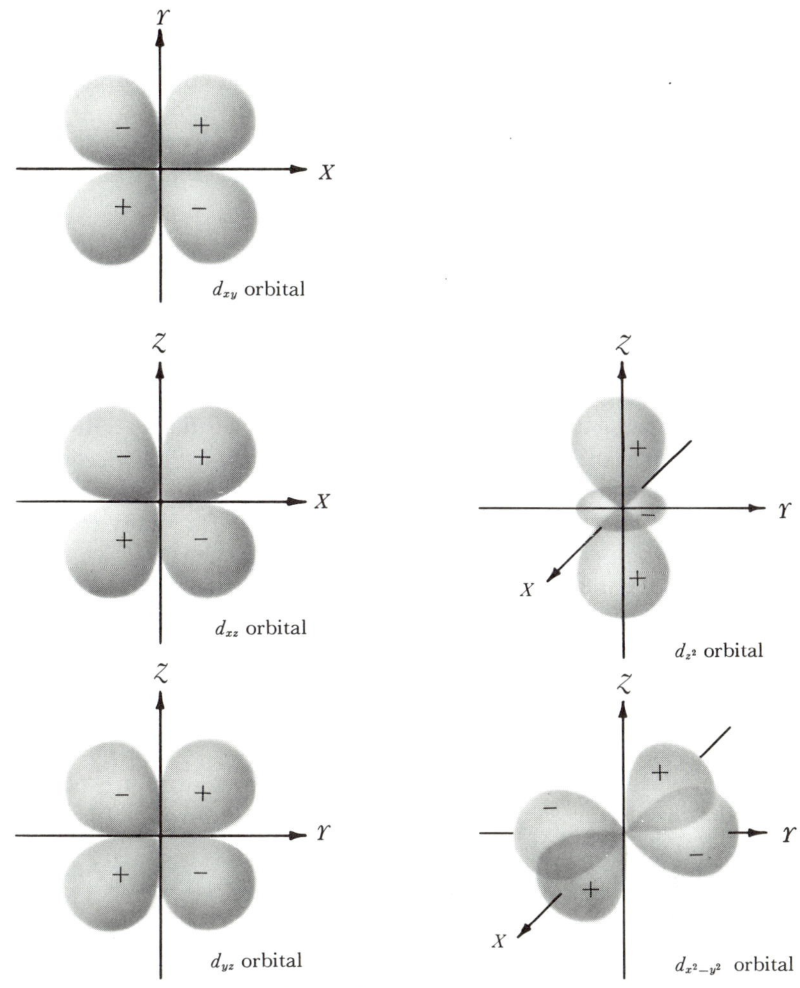

1–19
The five 3*d* atomic orbitals of hydrogen. The 4*d*, 5*d*, and 6*d* orbitals can be considered as essentially identical to these 3*d* orbitals except for an increase in size. Notice how the sign of the wave function changes from one lobe to the next in a given orbital. This change of sign is important when atomic orbitals are combined to make molecular orbitals.

1–14) and we usually represent all *p* orbitals by the three spatial pictures shown in Figure 1–18.

When $n = 3$ we encounter for the first time a value of two for l, thereby giving five possible values for the m_l quantum number. Thus there are five *d* orbitals. Although their mathematical representations are more complicated, pictorial representations of five energetically equivalent *d* orbitals are reasonably simple, as indicated in Figure 1–19. Notice that the d_{z^2} orbital, with $m_l = 0$, has a different shape than the others. Again, the signs labeling the various lobes indicate that the $3d$-orbital wave function is either positive or negative in that region. The d-orbital functions are symmetrical because inversion at the origin does not result in a change in algebraic sign.

There are orbitals of high l values in the shells with principal quantum numbers $n > 3$, but these orbitals are much more complicated and will not be introduced.

1–12 MANY-ELECTRON ATOMS

The Schrödinger equation can be set up for atoms with more than one electron, but it cannot be solved exactly in these cases. The second and subsequent electrons introduce the complicating feature of electron–electron repulsion, which is not present in the hydrogenlike atom. Nevertheless, with some modification the hydrogenlike orbitals account adequately for the electronic structures of many-electron atoms.

The key to the building or *Aufbau* process for many-electron atoms is called the *Pauli exclusion principle*, which states that no two electrons in an atom can have the same set of four quantum numbers. Thus two electrons in the ground state of atomic helium ($Z = 2$) must possess the following quantum numbers:

$$n = 1, \quad l = 0, \quad m_l = 0, \quad m_s = +\tfrac{1}{2}$$

and

$$n = 1, \quad l = 0, \quad m_l = 0, \quad m_s = -\tfrac{1}{2}$$

In other words, the two electrons in the helium atom are placed in the $1s$ orbital *with opposite spins* to be consistent with the Pauli principle. We abbreviate the orbital electronic structure of helium

$$\underset{1s}{(\uparrow\downarrow)} \quad \text{or} \quad 1s^2$$

in which $\uparrow$ stands for $m_s = +\frac{1}{2}$ and $\downarrow$ stands for $m_s = -\frac{1}{2}$.

In discussing the properties of many-electron atoms such as helium the concept of *effective nuclear charge* (Z_{eff}) is quite useful. Of course, the actual nuclear charge in a helium atom is +2. But the full force of this +2 charge is partially offset by the mutual repulsion of the two electrons. As far as either *one* of the electrons is concerned the Z_{eff} is *less* than +2, because of the "shielding" provided by the other electron. One atomic property that illustrates that Z_{eff} is less than +2 is the IE of helium, which is 24.59 eV. If there had been no electron–electron repulsion, each electron would have felt the full +2 nuclear charge. The IE in this hypothetical situation would have been equal to the value for a hydrogenlike atom with $Z = +2$, or $IE = (2)^2(13.6 \text{ eV}) = 54.4$ eV. Thus electron–electron repulsion reduces drastically the IE of helium, and the utility of the concept of Z_{eff} is established.

Another problem is encountered in the lithium atom, with $Z = 3$. The $1s$ orbital now is occupied fully by two electrons. The third electron must be placed in one of the orbitals with $n = 2$. But which one? The decision is not important in a hydrogenlike atom because $2s$ and $2p$ orbitals have the same energy. However, in a many-electron atom the shielding of any given electron from the nuclear charge by the other electrons depends on the l quantum number of the electron under consideration. For example, consider the $2s$ and $2p$ orbitals in the lithium atom. Both orbitals are shielded from the $+3$ nuclear charge by the $1s^2$ electrons, but the $2s$ orbital, because of its larger probability density very close to the nucleus (see Figure 1–15), is not shielded as strongly as is a $2p$ orbital. We say that the $2s$ orbital "penetrates" the inner $1s^2$ electron shell better than a $2p$ orbital does. Therefore the order of energies is $2s < 2p$, and the third electron in the lithium atom occupies the $2s$ orbital in the ground state:

(↑↓) (↑) or $1s^22s^1$
$1s$ $2s$

A beryllium atom ($Z = 4$) has a filled $2s$ orbital, and a boron atom ($Z = 5$) has the fifth electron in a $2p$ orbital:

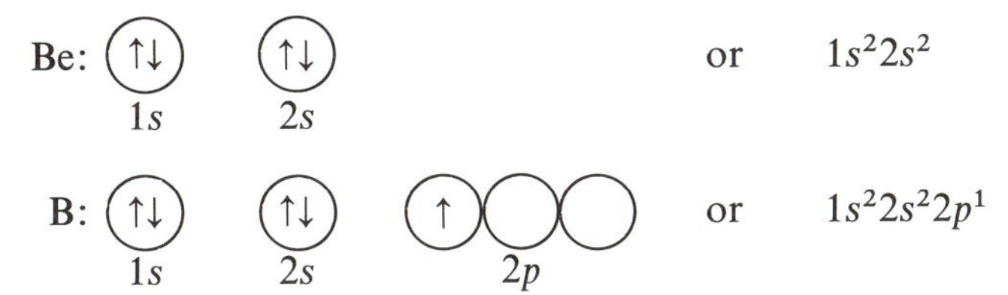

For a carbon atom ($Z = 6$) we have a choice of electron placement because there are a number of possible configurations for the second electron in a set of three $2p$ orbitals. For example, the three configurations

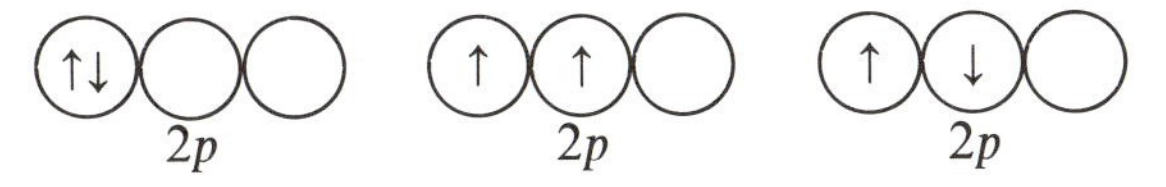

all obey the Pauli principle. Which configuration most accurately represents the ground state of atomic carbon? The choice is made by invoking *Hund's rule*, which states that for any set of orbitals of equal energy the electronic configuration with the maximum number of parallel spins results in the lowest electron–electron repulsion. Thus the ground-state configuration of atomic carbon is

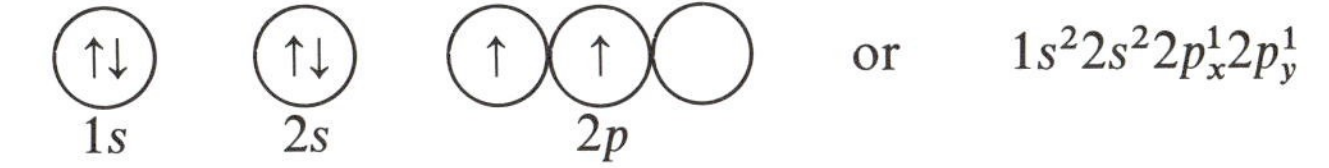

The two electrons with parallel spins (same m_s value) are said to be "unpaired."

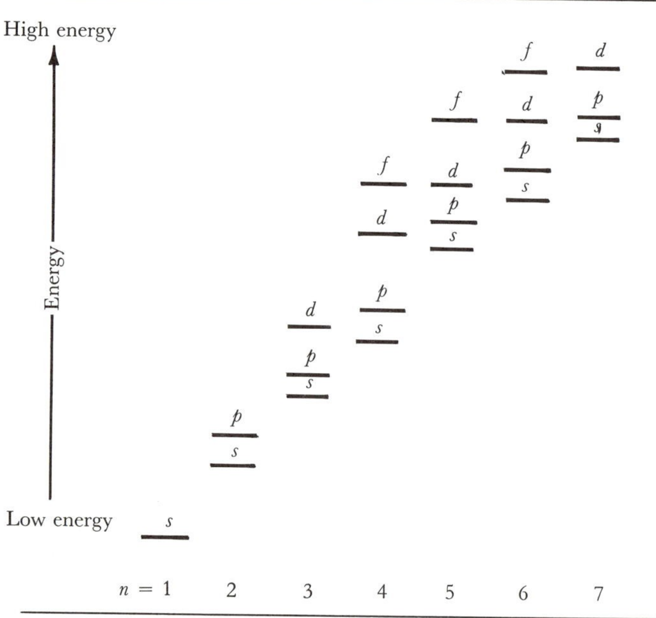

1–20
Relative energies of the orbitals in neutral, many-electron atoms. In building the periodic table for many-electron, neutral atoms, electrons are put in the available orbital of lowest energy. For example, the 4*s* orbital is filled before the 3*d* orbital, the 6*s* orbital is filled before the 4*f* or 5*d* orbitals, and so on.

Now we are in a position to build the ground-state configuration of the atoms of all elements by filling the atomic orbital sets with electrons in order of increasing energy, making certain that the Pauli principle and Hund's rule are obeyed. The total number of electrons that the different orbital sets can accommodate is given in Table 1–2. The *s*, *p*, *d*, and *f* orbital sets usually are called *subshells*. As we noted previously, the group of subshells for any given *n* value is called a shell.

Table 1–2. The *s*, *p*, *d*, and *f* orbital sets

Type of orbital	Orbital quantum numbers	Total orbitals in set	Total number of electrons that can be accommodated
s	$l = 0$; $m_l = 0$	1	2
p	$l = 1$; $m_l = 1, 0, -1$	3	6
d	$l = 2$; $m_l = 2, 1, 0, -1, -2$	5	10
f	$l = 3$; $m_l = 3, 2, 1, 0, -1, -2, -3$	7	14

We have discussed the fact that the $2p$ orbitals have higher energy than the $2s$ orbital in terms of different degrees of shielding in a many-electron atom. Actually, the energy order was known from atomic spectral experiments long before a theoretical rationale was available. We also know both from atomic theory and from experiments that the complete order of increasing energy of the orbital sets of interest in building the periodic table for many-electron, neutral atoms is $1s$, $2s$, $2p$, $3s$, $3p$, $4s$, $3d$, $4p$, $5s$, $4d$, $5p$, $6s$, $4f \cong 5d$, $6p$, $7s$, $5f \cong 6d$. This relative ordering is shown in Figure 1–20.

We must emphasize that Figure 1–20 gives the order of filling of orbitals only for neutral atoms. Orbital energies depend heavily on the atomic number and on the charge on the atom (ion), as is illustrated for the $3d$ and $4s$ orbital levels in the first transition series of elements. The $4s$ orbital is occupied before the $3d$ orbitals for the scandium (Sc) atom, but in the Sc^{2+} ion the $3d$ orbital is filled before the $4s$ orbital. This means that for the neutral and ionic species the $4s$ and $3d$ energy levels in scandium have different relative positions. Generally, the transition metal positive ions have an nd level that is of lower energy than the $(n + 1)s$ level. This energetic relationship will be important when we discuss the electronic structures of transition metal compounds in Chapter 5.

SUGGESTIONS FOR FURTHER READING

A. W. Adamson, "Domain Representations of Orbitals," *J. Chem. Ed.* **42,** 141 (1965).

R. S. Berry, Advisory Council on College Chemistry Resource Paper on "Atomic Orbitals," *J. Chem. Ed.* **43,** 283 (1966).

J. B. Birks (editor), *Rutherford at Manchester*, Benjamin, Menlo Park, Calif., 1963.

I. Cohen and T. Bustard, "Atomic Orbitals: Limitations and Variations," *J. Chem. Ed.* **43,** 187 (1966).

S. Devons, "Recollections of Rutherford and the Cavendish," *Physics Today* **24,** 38 (1971).

U. Fano and L. Fano, *Basic Physics of Atoms and Molecules*, Wiley, New York, 1959.

R. P. Feynman, R. B. Leighton, and M. Sands, *The Feynman Lectures on Physics*, Vol. III, Addison-Wesley, Reading, Mass., 1965.

W. Heisenberg, *The Physical Principles of Quantum Theory*, Dover, New York, 1930.

R. M. Hochstrasser, *Behavior of Electrons in Atoms*, Benjamin, Menlo Park, Calif., 1964.

R. C. Johnson and R. R. Rettew, "Shapes of Atoms," *J. Chem. Ed.* **42,** 145 (1965).

M. Karplus and R. N. Porter, *Atoms and Molecules: An Introduction for Students of Physical Chemistry*, Benjamin, Menlo Park, Calif., 1970.

E. A. Ogryzlo and G. B. Porter, "Contour Surfaces for Atomic and Molecular Orbitals," *J. Chem. Ed.* **40,** 256 (1963).

B. Perlmutter-Hayman, "The Graphical Representation of Hydrogen-Like Functions," *J. Chem. Ed.* **46,** 428 (1969).

R. E. Powell, "The Five Equivalent *d* Orbitals," *J. Chem. Ed.* **45,** 1 (1968).

G. Thomson, *The Atom*, Oxford Univ. Press, New York, 1962.

V. F. Weisskopf, "How Light Interacts with Matter," *Scientific American* (September, 1968).

QUESTIONS AND PROBLEMS

1. In each of the following statements choose *one* of the four possibilities, (a), (b), (c), or (d), that most accurately completes the statement. Read the statements carefully.

   A) Rutherford, Geiger, and Marsden performed experiments in which a beam of helium nuclei (α particles) was directed at a thin piece of gold foil. They found that the gold foil (a) severely deflected most of the particles of the beam directed at it; (b) deflected very few of the particles of the beam and deflected these only very slightly; (c) deflected most of the particles of the beam but deflected these only very slightly; (d) deflected very few of the particles of the beam but deflected these severely.

   B) From the results in (A) Rutherford concluded that (a) electrons are massive particles; (b) the positively charged parts of atoms are extremely small and extremely heavy particles; (c) the positively charged parts of atoms are moving with a velocity approaching that of light; (d) the diameter of an electron is approximately equal to the diameter of the nucleus.

   C) Which one of the following statements concerning the Bohr theory of the hydrogen atom is not true? The theory (a) successfully explained the observed emission and absorption spectra of the hydrogen atom; (b) requires that the greater the energy of the electron in the hydrogen atom the greater its velocity; (c) requires that the energy of the electron in the hydrogen atom can have only certain discrete values; (d) requires that the distance of the electron from the nucleus in the hydrogen atom can have only certain discrete values.

2. Consider two hydrogen atoms. The electron in the first hydrogen atom is in the $n = 1$ Bohr orbit. The electron in the second hydrogen atom is in the $n = 4$ orbit. (a) Which atom has the ground-state electronic configuration? (b) In which atom is the electron moving faster? (c) Which orbit has the larger radius? (d) Which atom has the lower potential energy? (e) Which atom has the higher ionization energy?

3. How much energy is required to ionize a hydrogen atom in which the electron occupies the $n = 5$ Bohr orbit?

4. Set up an expression for the wavelength of the radiation that would be emitted by a He^+ ion when it decays from an excited state having principal quantum number $n = 4$ to a lower excited state having $n = 3$. Your expression should give the wavelength as a function of m, e, h, π, and c only. Calculate the numerical value of the wavelength of the emitted radiation.

5. Calculate the wavelength of a photon of visible light with a frequency of $0.66 \times 10^{15}\ \text{sec}^{-1}$. What is the energy of the photon? What is the wave number?
6. Calculate the energy released when a hydrogen atom decays from the state having the principal quantum number three to the state having the principal quantum number two.
7. Calculate the frequency of the light emitted when a hydrogen atom decays as in Problem 6. Calculate the wave number of the light emitted in the decay of the hydrogen atom described.
8. The ground-state electronic configuration of lithium is $1s^2 2s^1$. When lithium is heated in a flame, bright red light is emitted. The red color is due to light emission at a wavelength of 6708 Å. Furthermore, no light emission is observed at longer wavelengths. (a) Suggest possible explanations for the emission at 6708 Å. [*Hint:* The $1s^2$ electrons are not involved.] (b) What is the frequency of the light? (c) What is the wave number (cm^{-1}) of the light? (d) What is the energy of the process in kilocalories per mole?
9. Following are several electronic configurations that may be correct for the nitrogen atom ($Z = 7$). Electrons are represented by arrows whose direction indicates the value of the spin quantum number, m_s. The three circles for the p orbitals indicate the possible values of the orbital-orientation quantum number, m_l. For each configuration write one of the following words: "excited," if the configuration represents a possible excited state of the nitrogen atom; "ground," if the configuration represents the ground state of the nitrogen atom; "forbidden," if the configuration in question cannot exist.

	$1s$	$2s$	$2p$	$3s$	$3p$
a)	(↑)	(↑↓)	(↑)(↑)(↑)	(↓)	( )( )( )
b)	(↑↑)	(↑↓)	(↑)(↑)(↑)	( )	( )( )( )
c)	(↓↑)	(↑↓)	(↓)(↓)(↓)	( )	( )( )( )
d)	(↓↑)	(↑↓)	(↓↑)( )(↑)	( )	( )( )( )
e)	(↑↓)	(↑)	( )(↑)( )	(↑)	(↑)(↑)( )
f)	(↑↓)	(↑↓)	(↑)(↓)(↑)	( )	( )( )( )

10. Write the orbital electronic structures for the following atoms and ions and, where appropriate, show that you know Hund's rule: P($Z = 15$); Na($Z = 11$); As($Z = 33$); C^- ($Z = 6$); O^+ ($Z = 8$).
11. Determine the number of unpaired electrons in the following atoms: C ($Z = 6$); F ($Z = 9$); Ne ($Z = 10$).
12. Draw spatial representations for the following orbitals and put in X, Y, and Z coordinates, if needed: $2p_z$; $3s$; $3d_{x^2-y^2}$; $3d_{xz}$; $n = 2, l = 1$.

13. An electron is in one of the $3d$ orbitals. What are the possible values of the orbital quantum numbers n, l, and m_l for the electron?
14. Write the orbital electronic structure of the ground state of (a) the calcium atom ($Z = 20$), and (b) the Mg^{2+} ion ($Z = 12$).
15. The Balmer series for atomic hydrogen occurs in the visible region of the spectrum. Which series in the emission spectrum of Be^{3+} has its lowest-energy line closest to the first line in the hydrogen Balmer series? How many quantum-number combinations are there for each of the participating energy levels for this emission line in Be^{3+}?
16. Suppose that you discovered some material from another universe that obeyed the following restrictions on quantum numbers:

$$n > 0$$
$$l + 1 \leq n$$
$$m_l = +l \quad \text{or} \quad -l$$
$$m_s = +\tfrac{1}{2}$$

Assume that Hund's rule still applies. What would be the atomic numbers of the first three noble gases in that universe? What would be the atomic numbers of the first three halogens?

17. The spectrum of He^+ contains, along with many others, lines at 329,170 cm^{-1}, 399,020 cm^{-1}, 411,460 cm^{-1}, and 421,330 cm^{-1}. Show that these lines fit a Rydberg-type equation, $\bar{\nu} = R(1/n_1^2 - 1/n_2^2)$. What is the ratio of R_H ($R_H = 109{,}678$ cm^{-1}) to R_{He^+}? Do these facts agree with Bohr's theory of hydrogenlike atoms?
18. An atom is observed to emit light at 1000 Å, 1250 Å, and 5000 Å. Theoretical considerations indicate that there are only two excited states involved. Explain why three lines are observed. How far in energy are the excited states above the ground state?
19. Why do the $4s$, $4p$, $4d$, and $4f$ orbitals have the same energy in the hydrogen atom but different energies in many-electron atoms?
20. How can the same atom of hydrogen, in quick succession, emit a photon in the Brackett, Paschen, Balmer, and Lyman series? Can it emit them in the reverse order? Why, or why not?

2

Electronic Properties of Atoms and Molecules

The buildup of orbital electronic structure guided by the Pauli principle shows clearly the simple basis for the periodic behavior of the elements. Generally, atoms with the same outer-orbital structure appear in the same column (group) of the periodic table. For example, atoms of the noble gas elements all have completely filled *ns* and *np* orbitals (closed-shell configurations). Metal atoms have very few electrons in the outermost *s* and *p* orbitals, thus they have a tendency to lose these electrons to achieve stable, closed-shell configurations. In contrast, the outer-orbital structures of nonmetals equal or exceed the s^2p^2 (halfway to s^2p^6) configuration. Nonmetal atoms sometimes gain electrons to achieve stable, closed-shell configurations.

The first-row transition metals are the elements scandium through zinc. These ten elements are the first to have orbital structures involving *d* electrons. From experience we know that the outer *s* and *p* electrons are the principal determinants of the chemical properties of atoms. Thus addition of the ten electrons to the 3*d* level does not alter grossly the chemical properties of these elements. The result is a "long period" of transition elements, all with similar properties.

2–1 LEWIS STRUCTURES FOR ATOMS

For all but a few atoms it is tedious to write the complete orbital electronic structure. It also is unnecessary because only the outer electrons are important in chemical reactions. We call the chemically important or outer electrons the *valence electrons*. The valence electrons of an atom are the electrons in the *s* and *p* orbitals beyond the closed-shell configurations. For example, in lithium the two 1*s* electrons are bound tightly to the nucleus of charge +3. Like the two electrons in helium, they are chemically unreactive. Thus we say that the valence electronic structure of lithium is $2s^1$, or Li·, in which the symbol Li represents the lithium nucleus and the two 1*s* electrons. The shorthand "dot" structure is called a *Lewis structure*, after G. N. Lewis. The Lewis notation vastly simplifies writing atomic structures.

Valence-orbital structures are so important in chemistry that all serious students should learn them for the main group elements. The learning task is made easy because the valence electronic structures are periodic. For example, oxygen, sulfur, and selenium atoms have the same valence structure ns^2np^4. Using sulfur as an example we write the Lewis formula for these atoms as $\cdot\ddot{\underset{\cdot}{S}}:$. If we know the valence electronic structure of atomic nitrogen is $\cdot\dot{\underset{\cdot}{N}}\cdot$, then we can write the Lewis formula for atomic phosphorus as $\cdot\dot{\underset{\cdot}{P}}\cdot$, because phosphorus is below nitrogen in the periodic table.

Now consider the Lewis structures for chlorine and chloride ion. The closed-shell structure before chlorine is the neon structure, $1s^22s^22p^6$, and chlorine has seven electrons in addition to this closed-shell configuration. Thus the Lewis atomic structure for chlorine has seven dots, :Ċ̣l:, in which Cl represents the nucleus and the $1s^2 2s^2 2p^6$ electrons. When one electron is added to a chlorine atom to produce a chloride ion there are eight valence electrons, thereby giving the closed-shell configuration. The chloride ion has the structure :C̤̈l:$^-$, which shows the charge of -1.

2–2 EFFECTIVE ATOMIC RADII IN MOLECULES

Now we turn our attention to the relationship between atomic properties and valence-orbital structures. First we will consider the effective radius of an atom in a molecule. The *effective atomic radius* of an atom is defined as one half the distance between two nuclei of the element that are held together by a purely covalent single bond. (A covalent bond is a pair of electrons shared between two atoms.) For example, the separation of the two protons in the hydrogen molecule, H_2, is 0.74 Å. Thus we assign each hydrogen atom in the H_2 molecule an atomic radius of 0.37 Å. The distance between lithium nuclei in Li_2 is 2.67 Å. Thus the atomic radius of lithium is approximately 1.34 Å. The average effective radii of atoms of a selection of representative elements shown in the periodic-table arrangement of Figure 2–1 were determined from experimentally observed bond distances in many molecules. The atomic radius in most cases is compared with the size of the appropriate closed-shell positive or negative ion.

In terms of orbital structure the explanation of the shrinkage of atomic radii across a given row (or period) in Figure 2–1 is as follows. In any given period electrons are added to *s* and *p* orbitals, which are not able to shield each other effectively from the increasing positive nuclear charge. Thus an increase in the positive charge of the nucleus results in an increase in the effective nuclear charge, Z_{eff}, thereby decreasing the effective atomic radius. This is the reason why a beryllium atom, for example, is smaller than a lithium atom.

From hydrogen to lithium there is a large increase in effective atomic radius. The reason is that a third electron in a lithium atom is in an orbital that has a much larger effective radius than the hydrogen 1*s* orbital. According to the Pauli principle the third electron in lithium must be in an orbital with a larger principal quantum number, namely the 2*s* orbital. Seven more electrons can be added to the 2*s* and 2*p* orbitals, which have approximately the same radii. However, these electrons do not effectively shield each other from the positive nuclear charge as it increases, and the result is an increase in Z_{eff} and a corresponding decrease in radii in the series lithium ($Z = 3$) through neon ($Z = 10$). After neon additional electrons cannot be accommodated by the $n = 2$ level. Thus an eleventh electron must go into the $n = 3$ level, specifically, into the 3*s* orbital. Since the effective radii increase from the $n = 1$ to $n = 2$ to $n = 3$ valence orbitals, the effective size of an atom also increases with increasing atomic number within each group in the periodic table.

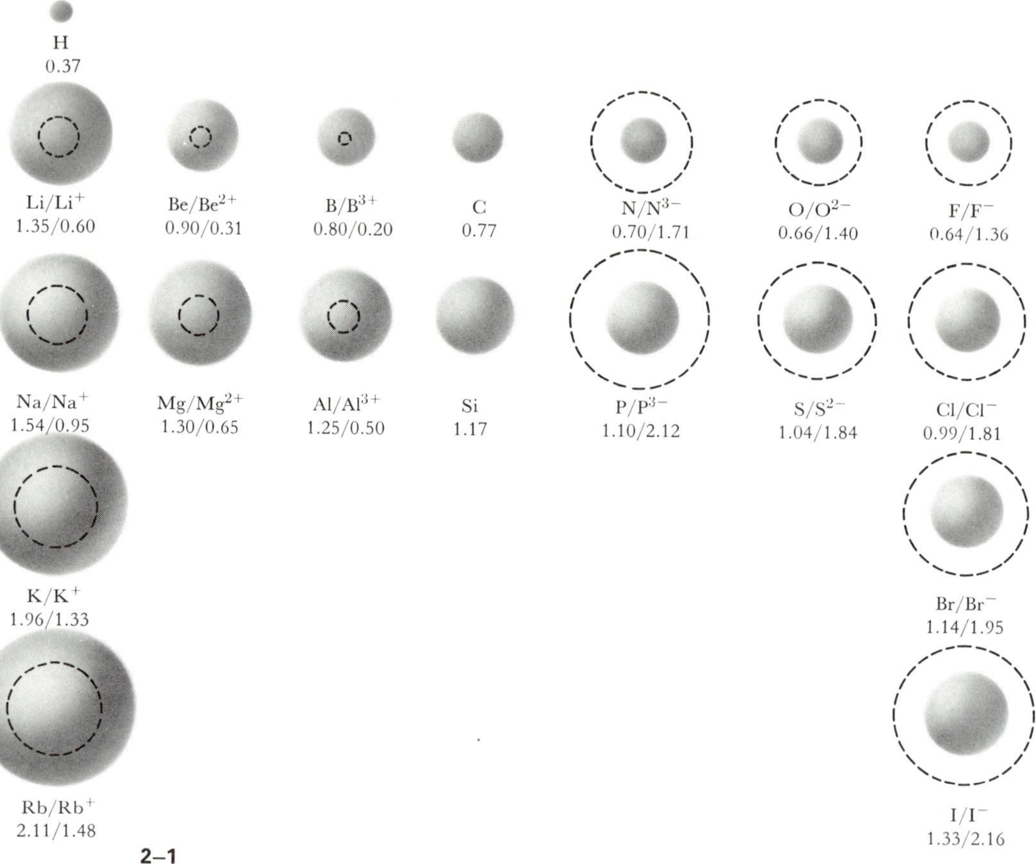

2–1
Relative atomic radii of some elements compared with the radii of the appropriate closed-shell ions. Radii are in Å. Solid spheres represent atoms and dashed circles represent ions. Notice that positive ions are smaller than their neutral atoms and negative ions are larger. Why should this be so?

2–3 IONIZATION ENERGIES AND ORBITAL CONFIGURATIONS

As we discussed in Section 1–6, the ionization energy, IE, of an atom is the energy required to remove an electron from the gaseous atom. The first ionization energy, IE_1, is the energy needed to remove one electron from the neutral gaseous atom to produce a unipositive gaseous ion. The process can be written

$$\text{atom}(g) + (\text{energy} = IE_1) \rightarrow \text{ion}^+(g) + e^-$$

For sodium we write

$$\text{Na}(g) + (IE_1 = 5.139 \text{ eV}) \rightarrow \text{Na}^+(g) + e^-$$

2–2
Average single-bond atomic radius of the lithium atom compared with the average ionic radius of Li^+ and the calculated Bohr radius of Li^{2+}.

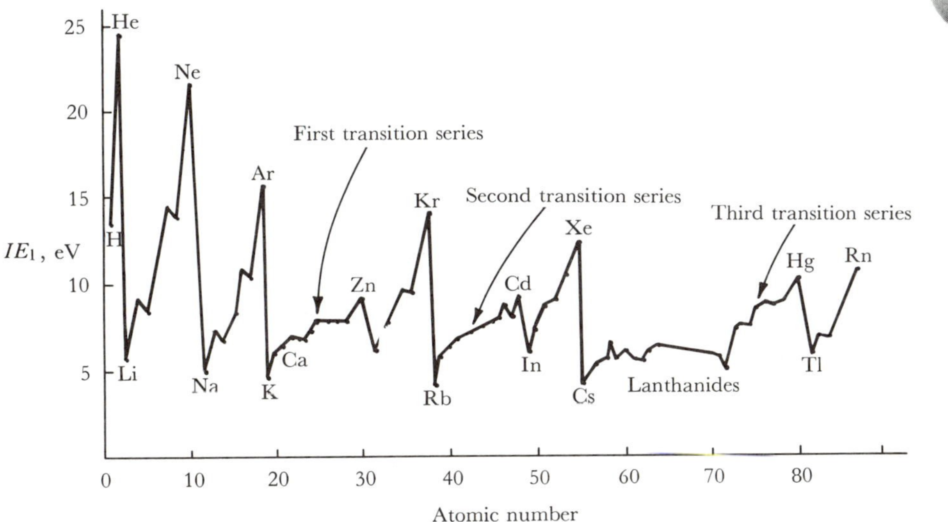

2–3
Variation of atomic ionization energy, in electron volts, with atomic number. Notice that maximum ionization energies in a given row occur for the noble gases and that the ionization energies of the transition elements are similar.

The first ionization energies for most of the elements are listed in Table 2–1, together with their ground-state orbital electronic configurations. For any atom IE_1 is always the smallest ionization energy.

In all atoms but hydrogen further ionizations are possible. For example, the three ionizations for lithium are

$$\mathrm{Li}(g) + (IE_1 = 5.392\ \mathrm{eV}) \rightarrow \mathrm{Li}^+(g) + e^-$$
$$\mathrm{Li}^+(g) + (IE_2 = 75.638\ \mathrm{eV}) \rightarrow \mathrm{Li}^{2+}(g) + e^-$$
$$\mathrm{Li}^{2+}(g) + (IE_3 = 122.45\ \mathrm{eV}) \rightarrow \mathrm{Li}^{3+}(g) + e^-$$

The large increase in the order $IE_1 < IE_2 < \cdots < IE_n$ is understandable because as electrons are lost the effective nuclear charge, Z_{eff}, increases. As a result the effective radius of an atom or ion decreases sharply and the net attraction between the electrons and Z_{eff} increases sharply in a series A, A^+, A^{2+}, A^{3+}, $\cdots$. The relative effective sizes of Li, Li^+, and Li^{2+} are shown in Figure 2–2.

Table 2–1. Electronic configuration and ionization energies

Z	Atom	Orbital electronic configuration	IE_1, electron volts	IE_2, electron volts
1	H	$1s^1$	13.598	—
2	He	$1s^2$	24.587	54.416
3	Li	(He)$2s^1$	5.392	75.638
4	Be	(He)$2s^2$	9.322	18.211
5	B	(He)$2s^2 2p^1$	8.298	25.154
6	C	(He)$2s^2 2p^2$	11.260	24.383
7	N	(He)$2s^2 2p^3$	14.534	29.601
8	O	(He)$2s^2 2p^4$	13.618	35.116
9	F	(He)$2s^2 2p^5$	17.422	34.970
10	Ne	(He)$2s^2 2p^6$	21.564	40.962
11	Na	(Ne)$3s^1$	5.139	47.286
12	Mg	(Ne)$3s^2$	7.646	15.035
13	Al	(Ne)$3s^2 3p^1$	5.986	18.828
14	Si	(Ne)$3s^2 3p^2$	8.151	16.345
15	P	(Ne)$3s^2 3p^3$	10.486	19.725
16	S	(Ne)$3s^2 3p^4$	10.360	23.33
17	Cl	(Ne)$3s^2 3p^5$	12.967	23.81
18	Ar	(Ne)$3s^2 3p^6$	15.759	27.629
19	K	(Ar)$4s^1$	4.341	31.625
20	Ca	(Ar)$4s^2$	6.113	11.871
21	Sc	(Ar)$4s^2 3d^1$	6.54	12.80
22	Ti	(Ar)$4s^2 3d^2$	6.82	13.58
23	V	(Ar)$4s^2 3d^3$	6.74	14.65
24	Cr	(Ar)$4s^1 3d^5$	6.766	16.50
25	Mn	(Ar)$4s^2 3d^5$	7.435	15.640
26	Fe	(Ar)$4s^2 3d^6$	7.870	16.18
27	Co	(Ar)$4s^2 3d^7$	7.86	17.06
28	Ni	(Ar)$4s^2 3d^8$	7.635	18.168
29	Cu	(Ar)$4s^1 3d^{10}$	7.726	20.292
30	Zn	(Ar)$4s^2 3d^{10}$	9.394	17.964
31	Ga	(Ar)$4s^2 3d^{10} 4p^1$	5.999	20.51
32	Ge	(Ar)$4s^2 3d^{10} 4p^2$	7.899	15.934
33	As	(Ar)$4s^2 3d^{10} 4p^3$	9.81	18.633
34	Se	(Ar)$4s^2 3d^{10} 4p^4$	9.752	21.19
35	Br	(Ar)$4s^2 3d^{10} 4p^5$	11.814	21.8
36	Kr	(Ar)$4s^2 3d^{10} 4p^6$	13.999	24.359
37	Rb	(Kr)$5s^1$	4.177	27.28
38	Sr	(Kr)$5s^2$	5.695	11.030
39	Y	(Kr)$5s^2 4d^1$	6.38	12.24
40	Zr	(Kr)$5s^2 4d^2$	6.84	13.13
41	Nb	(Kr)$5s^1 4d^4$	6.88	14.32
42	Mo	(Kr)$5s^1 4d^5$	7.099	16.15
43	Tc	(Kr)$5s^2 4d^5$	7.28	15.26
44	Ru	(Kr)$5s^1 4d^7$	7.37	16.76
45	Rh	(Kr)$5s^1 4d^8$	7.46	18.08
46	Pd	(Kr)$4d^{10}$	8.34	19.43
47	Ag	(Kr)$5s^1 4d^{10}$	7.576	21.49
48	Cd	(Kr)$5s^2 4d^{10}$	8.993	16.908
49	In	(Kr)$5s^2 4d^{10} 5p^1$	5.786	18.869
50	Sn	(Kr)$5s^2 4d^{10} 5p^2$	7.344	14.632
51	Sb	(Kr)$5s^2 4d^{10} 5p^3$	8.641	16.53
52	Te	(Kr)$5s^2 4d^{10} 5p^4$	9.009	18.6

Table 2–1 (*Continued*)

Z	Atom	Orbital electronic configuration	IE_1, electron volts	IE_2, electron volts
53	I	(Kr)$5s^24d^{10}5p^5$	10.451	19.131
54	Xe	(Kr)$5s^24d^{10}5p^6$	12.130	21.21
55	Cs	(Xe)$6s^1$	3.894	25.1
56	Ba	(Xe)$6s^2$	5.212	10.004
57	La	(Xe)$6s^25d^1$	5.577	11.06
58	Ce	(Xe)$6s^24f^15d^1$	5.47	10.85
59	Pr	(Xe)$6s^24f^3$	5.42	10.55
60	Nd	(Xe)$6s^24f^4$	5.49	10.72
61	Pm	(Xe)$6s^24f^5$	5.55	10.90
62	Sm	(Xe)$6s^24f^6$	5.63	11.07
63	Eu	(Xe)$6s^24f^7$	5.67	11.25
64	Gd	(Xe)$6s^24f^75d^1$	6.14	12.1
65	Tb	(Xe)$6s^24f^9$	5.85	11.52
66	Dy	(Xe)$6s^24f^{10}$	5.93	11.67
67	Ho	(Xe)$6s^24f^{11}$	6.02	11.80
68	Er	(Xe)$6s^24f^{12}$	6.10	11.93
69	Tm	(Xe)$6s^24f^{13}$	6.18	12.05
70	Yb	(Xe)$6s^24f^{14}$	6.254	12.17
71	Lu	(Xe)$6s^24f^{14}5d^1$	5.426	13.9
72	Hf	(Xe)$6s^24f^{14}5d^2$	7.0	14.9
73	Ta	(Xe)$6s^24f^{14}5d^3$	7.89	—
74	W	(Xe)$6s^24f^{14}5d^4$	7.98	—
75	Re	(Xe)$6s^24f^{14}5d^5$	7.88	—
76	Os	(Xe)$6s^24f^{14}5d^6$	8.7	—
77	Ir	(Xe)$6s^24f^{14}5d^7$	9.1	—
78	Pt	(Xe)$6s^14f^{14}5d^9$	9.0	18.563
79	Au	(Xe)$6s^14f^{14}5d^{10}$	9.225	20.5
80	Hg	(Xe)$6s^24f^{14}5d^{10}$	10.437	18.756
81	Tl	(Xe)$6s^24f^{14}5d^{10}6p^1$	6.108	20.428
82	Pb	(Xe)$6s^24f^{14}5d^{10}6p^2$	7.416	15.032
83	Bi	(Xe)$6s^24f^{14}5d^{10}6p^3$	7.289	16.69
84	Po	(Xe)$6s^24f^{14}5d^{10}6p^4$	8.42	—
85	At	(Xe)$6s^24f^{14}5d^{10}6p^5$	—	—
86	Rn	(Xe)$6s^24f^{14}5d^{10}6p^6$	10.748	—
87	Fr	(Rn)$7s^1$	—	—
88	Ra	(Rn)$7s^2$	5.279	10.147
89	Ac	(Rn)$7s^26d^1$	6.9	12.1
90	Th	(Rn)$7s^26d^2$	—	11.5
91	Pa	(Rn)$7s^25f^26d^1$	—	—
92	U	(Rn)$7s^25f^36d^1$	—	—
93	Np	(Rn)$7s^25f^46d^1$	—	—
94	Pu	(Rn)$7s^25f^6$	5.8	—
95	Am	(Rn)$7s^25f^7$	6.0	—
96	Cm	(Rn)$7s^25f^76d^1$	—	—
97	Bk	(Rn)$7s^25f^9$	—	—
98	Cf	(Rn)$7s^25f^{10}$	—	—
99	Es	(Rn)$7s^25f^{11}$	—	—
100	Fm	(Rn)$7s^25f^{12}$	—	—
101	Md	(Rn)$7s^25f^{13}$	—	—
102	No	(Rn)$7s^25f^{14}$	—	—
103	Lr	(Rn)$7s^25f^{14}6d^1$	—	—

Table 2–2. Periodic behavior of ionization energies (*IE*)

General increase of *IE*, electron volts ⟶								
Decrease of *IE*, electron volts ↓	H· 13.60							He: 24.59
	Li· 5.39	Be. 9.32	·B· 8.30	·C· 11.26	·N· 14.53	·O: 13.62	:F: 17.42	:Ne: 21.56
	Na· 5.14							

Ionization energies of atoms exhibit periodic behavior, as illustrated for IE_1 in Figure 2–3. We see that in any given row in the periodic table the ionization energies generally increase with increasing atomic number, being smallest for the lithium family and largest for the helium family. However, there are irregularities; atoms with filled or half-filled subshells have larger ionization energies than might be expected. For example, beryllium ($2s^2$) and nitrogen ($2s^2 2p^3$) have larger ionization energies than boron and oxygen atoms, respectively (Table 2–2).

The increase (although slightly irregular) in *IE*'s from lithium to neon is due to the steady increase in Z_{eff} with increasing atomic number. From lithium ($Z = 3$) to neon ($Z = 10$) all valence electrons are accommodated in $2s$ and $2p$ orbitals and are not able to shield each other completely from the increasing nuclear charge. Notice that the completely filled orbital structures in the helium family are especially stable. Electrons in helium and neon, which have the compact $1s^2$ and $2s^2 2p^6$ structures, respectively, are attracted relatively closely to the nucleus, thus the energies needed to remove an electron from these noble-gas atoms are correspondingly large.

Figure 2–3 also shows that in any given family of elements the ionization energies decrease with increasing atomic number. For example, the IE_1 of sodium is less than the IE_1 of lithium. Recall that the $3s$ valence orbital in sodium has a larger effective radius than the $2s$ valence orbital in lithium. According to Coulomb's law the net attraction between the $3s$ electron in a sodium atom and its effective nuclear charge is less than the net attraction between the $2s$ electron in a lithium atom and its effective nuclear charge. Thus it takes less energy to remove the $3s$ electron from the sodium atom than it does to remove the $2s$ electron from the lithium atom.

2–4 ELECTRON AFFINITY

The *electron affinity*, *EA*, of an atom is the energy change accompanying the addition of one electron to a neutral gaseous atom to produce a negative ion. Electron affinity is defined by the equation

$$\text{atom}(g) + e^- \rightarrow \text{ion}^-(g) + (\text{energy} = EA)$$

Thus if energy is released when an atom acquires one electron the atomic *EA* is positive. If energy is required for the reaction *EA* is negative. There are serious experimental problems in determining accurate *EA* values. As a result only a few *EA* values are known precisely. The better known electron affinities are listed in Table 2–3.

Table 2–3. Atomic electron affinities (*EA*)

Atom	Orbital electronic configuration	*EA*, electron volts	Orbital electronic configuration of anion
H	$1s^1$	0.756	(He)
F	(He)$2s^2 2p^5$	3.45	(Ne)
Cl	(Ne)$3s^2 3p^5$	3.61	(Ar)
Br	(Ar)$4s^2 3d^{10} 4p^5$	3.36	(Kr)
I	(Kr)$5s^2 4d^{10} 5p^5$	3.06	(Xe)
O	(He)$2s^2 2p^4$	1.47	(He)$2s^2 2p^5$
S	(Ne)$3s^2 3p^4$	2.07	(Ne)$3s^2 3p^5$
Se	(Ar)$4s^2 3d^{10} 4p^4$	(1.7)	(Ar)$4s^2 3d^{10} 4p^5$
Te	(Kr)$5s^2 4d^{10} 5p^4$	(2.2)	(Kr)$5s^2 4d^{10} 5p^5$
N	(He)$2s^2 2p^3$	(−0.1)	(He)$2s^2 2p^4$
P	(Ne)$3s^2 3p^3$	(0.78)	(Ne)$3s^2 3p^4$
As	(Ar)$4s^2 3d^{10} 4p^3$	(0.6)	(Ar)$4s^2 3d^{10} 4p^4$
C	(He)$2s^2 2p^2$	1.25	(He)$2s^2 2p^3$
Si	(Ne)$3s^2 3p^2$	(1.39)	(Ne)$3s^2 3p^3$
Ge	(Ar)$4s^2 3d^{10} 4p^2$	(1.2)	(Ar)$4s^2 3d^{10} 4p^3$
B	(He)$2s^2 2p^1$	(0.3)	(He)$2s^2 2p^2$
Al	(Ne)$3s^2 3p^1$	(0.5)	(Ne)$3s^2 3p^2$
Ga	(Ar)$4s^2 3d^{10} 4p^1$	(0.18)	(Ar)$4s^2 3d^{10} 4p^2$
In	(Kr)$5s^2 4d^{10} 5p^1$	(0.2)	(Kr)$5s^2 4d^{10} 5p^2$
Be	(He)$2s^2$	(−0.6)	(He)$2s^2 2p^1$
Mg	(Ne)$3s^2$	(−0.3)	(Ne)$3s^2 3p^1$
Li	(He)$2s^1$	0.6	(He)$2s^2$
Na	(Ne)$3s^1$	(0.54)	(Ne)$3s^2$
Zn	(Ar)$4s^2 3d^{10}$	(−0.9)	(Ar)$4s^2 3d^{10} 4p^1$
Cd	(Kr)$5s^2 4d^{10}$	(−0.6)	(Kr)$5s^2 4d^{10} 5p^1$

Atoms in the fluorine family have relatively large electron affinities because an electron can be added to a valence *p* orbital relatively easily, thereby completing a closed-shell s^2p^6 configuration. Atoms that already have closed shells or subshells often have negative *EA* values. Examples are beryllium, magnesium, and zinc. Also, the nitrogen-family atoms, which have half-filled valence *p* subshells, have negative or very small positive *EA* values. Thus from the trends in both ionization energies and electron affinities it is apparent that a half-filled subshell is a particularly stable electronic arrangement.

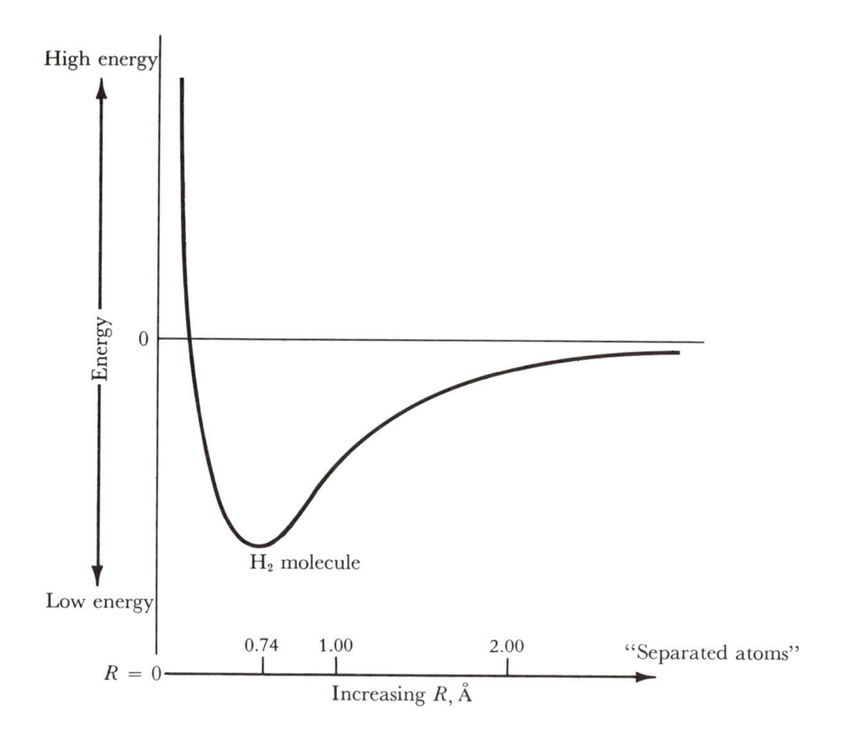

2–4
Potential energy curve for the H_2 molecule. As the distance between nuclei decreases the potential energy decreases because of electron–nucleus attraction, and then increases again because of nucleus–nucleus repulsion. The point of minimum energy corresponds to the equilibrium internuclear separation, or bond distance of H_2, which is 0.74 Å.

2–5 COVALENT BONDING

A covalent bond forms when atoms that have electrons of similar, or equal, valence-orbital energies combine. For example, two atoms of hydrogen are joined by a covalent bond in the H_2 molecule. To obtain some understanding of covalent bond formation we shall consider in detail the energy changes that occur if we allow two hydrogen atoms to come together from a large distance.

Each hydrogen atom consists of one electron and one proton. We will assume that the energy in the system is zero when the hydrogen atoms are isolated, because we want to focus our attention on the change in the energy of the system as we bring the two hydrogen atoms together. In each isolated atom there is a single important force of attraction, which is the force between the electron and the proton. However, if we bring two hydrogen atoms together there are additional attractive and repulsive forces that we must consider. Two such forces are the attractions of the first electron for the second nucleus and the second electron for the first nucleus. There also are repulsive forces: The

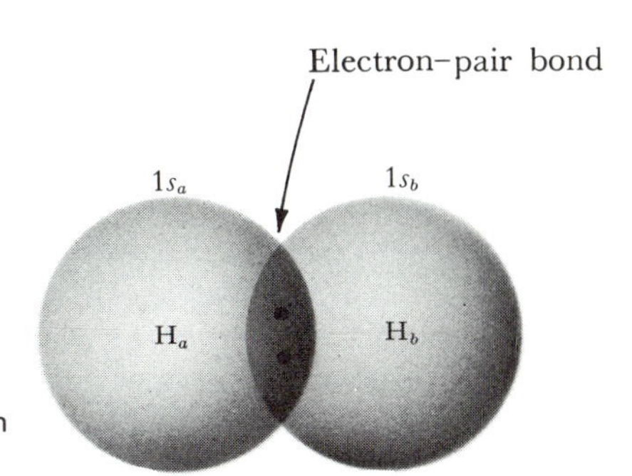

2–5
The two hydrogen 1*s* orbitals overlap to form an electron-pair covalent bond in H_2.

two electrons repel each other and the two protons repel each other. Thus there are four new electrostatic forces, two of them are attractive and two are repulsive.

The most important force as the two hydrogen atoms come together is the force of attraction between the two electrons and the two protons. As the hydrogen atoms are brought together the energy of the system decreases, and it continues to decrease until the two hydrogen atoms are so close together that the nuclear repulsion—the repulsion between the protons—becomes significant, thereby causing the energy to increase. Figure 2–4 shows the energy curve for this process. Notice that there is a point—an equilibrium internuclear separation—at which the two hydrogen atoms are bound together in a stable configuration and the energy of the system is a minimum. The separation of hydrogen nuclei at the position of minimum energy is the *equilibrium internuclear separation*, or the *bond distance* of H_2, which is 0.74 Å.

To separate two atoms in a diatomic molecule (e.g., H_2) requires approximately the energy difference between the minimum potential energy of the system and the zero energy of the isolated atoms. This is called the *bond energy* of a diatomic molecule. The bond energy usually is expressed in kilocalories per mole, that is, the number of kilocalories required to break one mole of bonds. The bond energy of H_2 is 103 kcal mole^{-1}.

The two electrons in H_2 are shared equally by the two hydrogen 1*s* orbitals. This, in effect, gives each hydrogen atom a stable, closed-shell (He-type) configuration. A simple orbital representation of the electron-pair bond in H_2 is shown in Figure 2–5.

Hydrogen molecule-ion

Removal of one electron from H_2 during electron bombardment of hydrogen at very low pressures forms the transient species H_2^+, which is the simplest of all molecules. This species, which is a combination of a proton and a hydrogen atom, is called the *hydrogen molecule-ion*. If a proton and a hydrogen atom come together, there is less net attractive force than when two hydrogen atoms approach each other. As a result the system's potential energy does not decrease as much, and the minimum energy occurs at a longer internuclear distance than for H_2. If we define one covalent bond as having two net bonding electrons, then H_2^+, with one electron, has one half a bond. We find that H_2, with two electrons, has a bond length of 0.74 Å and that H_2^+, with one electron, has a bond length of 1.06 Å. The bond energy for H_2 is 103 kcal mole^{-1} and the bond energy for H_2^+ is 61 kcal mole^{-1}. These comparisons illustrate the general fact that the more bonds there are between two atoms, the shorter the bond length and the stronger the bond. The number of bonds between two atoms is called the *bond order*.

2–6 PROPERTIES OF H_2 AND H_2^+ IN A MAGNETIC FIELD

Most substances can be classified as either paramagnetic or diamagnetic, depending on their behavior in a magnetic field. A *paramagnetic substance* is attracted to a magnetic field. A *diamagnetic substance* is repelled by a magnetic field. Generally, atoms and molecules with unpaired electrons are paramagnetic because there is a permanent magnetic moment associated with net electron spin. In many cases there is a further contribution to the permanent magnetic moment as a result of the movement of an electron in its orbital around the nucleus (or nuclei, in the case of molecules). In addition to the paramagnetic moment, magnetic moments are induced in atoms and molecules by applying an external magnetic field. Such induced moments are opposite to the direction of the external magnetic field, thus repulsion occurs. The magnitude of this repulsion is a measure of the diamagnetism of an atom or molecule.

The paramagnetism of atoms and small molecules that results from unpaired electrons is larger than induced diamagnetism, thus such substances are attracted to a magnetic field. Atoms and molecules with no unpaired electrons, therefore having no paramagnetism due to electron spin, are diamagnetic and are repelled by a magnetic field. The H_2^+ ion, with one unpaired electron, is paramagnetic. The H_2 molecule, with two paired electrons, is diamagnetic.

2–7 LEWIS STRUCTURES FOR DIATOMIC MOLECULES

The Lewis structures for two H atoms can be combined to give H:H, or H—H, in which the electron-pair bond is abbreviated with a single line. Lewis suggested that the stability of H_2 is due to the tendency of each hydrogen atom to associate itself with two electrons, thereby achieving a closed-shell configuration (He:). The Lewis structure of a fluorine atom is $:\ddot{\underset{..}{F}}\cdot$. Each fluorine atom lacks one electron to complete a closed-shell configuration. In the Lewis model two fluorine atoms achieve a closed-shell configuration by sharing two electrons, thereby forming an electron-pair bond ($:\ddot{\underset{..}{F}}—\ddot{\underset{..}{F}}:$).

The bond energy of F_2 is about 33 kcal mole^{-1}, compared to 103 kcal mole^{-1} for H_2. Thus it takes much less energy to separate two fluorine atoms in a fluorine molecule than it does to separate two hydrogen atoms in a hydrogen molecule. The reason may be due to the repulsions of the unshared or lone-pair electrons that are not involved in the bond in a fluorine molecule. This type of interaction is not present in a hydrogen molecule because there are no unshared electron pairs:

H—H	:F̤̈—F̤̈:
no unshared pairs	six unshared pairs

The H_2 and F_2 molecules are representative of many molecules in which electron-pair bonds are formed such that each atom achieves a closed-shell configuration. For hydrogen to achieve a closed-shell requires two electrons, which is the capacity of the 1*s* valence orbital. Each atom in the second row of the periodic table requires eight electrons (an octet) to achieve a closed shell because the 2*s* and 2*p* orbitals have a total capacity for eight electrons. This requirement commonly is known as the *octet rule*. In the example of the fluorine molecule, after bonding each fluorine atom has eight electrons associated with it. For third-row atoms eight electrons usually are associated with each atom because of their 3*s* and 3*p* valence orbitals. However, these atoms also have 3*d* orbitals (although their energy is substantially higher than that of the 3*p* orbitals), so more than eight electrons may be associated with atoms in the third and higher rows.

2–8 IONIC BONDING

In a pure covalent bond (e.g., in H_2) electrons are shared equally between two atoms. A pure ionic bond is the other extreme; that is, there is complete transfer of electrons from one atom to another and no sharing. There probably is no diatomic molecule that has a completely ionic bond. However, alkali-metal halide molecules have such unequal sharing of electrons that they may serve as models for ionic bonding.

Consider a sodium chloride diatomic molecule. If two atoms of different elements are involved in a two-electron bond, there must be unequal sharing of electrons. Unequal sharing is caused by differences in the ionization energies and electron affinities of the two atoms. In NaCl the sodium atom has a small ionization energy of about 5 eV and a small electron affinity of about 0.5 eV. Therefore it easily loses an electron to form Na^+. The chlorine atom has a large ionization energy of more than 10 eV and a large electron affinity of almost 4 eV. Thus it does not easily lose an electron, rather it easily can gain an electron. The preceding data provide a quantitative measure of the tendency for Cl to add an electron and for Na to lose an electron to achieve closed-shell configurations. Thus a sodium atom and a chlorine atom can be assumed to form an ionic bond in which the one 3*s* valence electron in sodium is transferred to the one vacancy in the chlorine 3*p* orbitals.

We consider all ionic molecules as being composed of interacting ions. Thus sodium chloride contains the Na^+ ion, which has the inert neon configuration, and the Cl^- ion, which has the inert argon configuration. Therefore the correct Lewis structure for NaCl is

$$[Na^+][:\ddot{\underset{..}{Cl}}:^-]$$

Ionic or "ion-pair" structures are reasonable models of bonds involving alkali and alkaline-earth metals with oxygen or one of the halogens. Ionic bonds are appropriate in these cases because the large differences in ionization energies and electron affinities lead to extremely unequal sharing of electrons.

A relatively simple calculation of the bond energy of sodium chloride will demonstrate that the ion-pair model is reasonable. We represent sodium chloride as a Na^+ ion and a Cl^- ion, separated by a bond distance R of 2.36 Å. Since from Coulomb's law an energy of 332 kcal mole^{-1} is required to dissociate completely oppositely charged bodies (each with unit charge) from a distance of 1 Å, the calculated energy for the process $NaCl \rightarrow Na^+ + Cl^-$ is 332/2.36 or 140 kcal mole^{-1}.

The *standard bond energy* (or bond dissociation energy) is the energy required to dissociate a molecule into its component atoms. For sodium chloride the process is $NaCl \rightarrow Na + Cl$. To find the energy required to dissociate NaCl into atoms we must add an electron to the Na^+ ion, which releases the atomic sodium ionization energy, and remove one electron from the Cl^- ion, which requires the atomic chlorine electron affinity energy. To complete the calculation we must convert the ionization energy and electron affinity from electron volts to kilocalories per mole through the relationship 1 eV = 23.069 kcal mole^{-1}. The molar ionization energy of sodium is 5.14 eV (119 kcal mole^{-1}) and the molar electron affinity of chlorine is 3.61 eV (83 kcal mole^{-1}). Thus in the transformation from ions to atoms 36 kcal mole^{-1} is gained (119 − 83), so the total process $Na^+Cl^- \rightarrow Na + Cl$ requires 104 kcal mole^{-1} (140 − 36). Therefore 104 kcal mole^{-1} is the calculated bond dissociation energy. The experimental value is 98 kcal mole^{-1}, so the ion-pair approximation allows us to calculate the bond dissociation energy within 6% of the experimental value.

Using the "point-charge" model we ignored the repulsions between filled electron shells. A better calculation using the pure ionic model, but taking the repulsion of the closed shells into account, gives the bond energy of NaCl within 2% of the experimental value. This and similar calculations indicate that an ionic model is justified for molecules formed from lithium- or beryllium-family atoms and either oxygen or fluorine-family atoms. In addition, direct experimental evidence from electron-diffraction data shows that the electron distribution in a sodium chloride molecule is approximately that given by the ionic model.

2–9 ELECTRONEGATIVITY

Between the extremes of covalent and ionic bonds there are intermediate cases for which neither bond type is a sufficient description. A discussion of these intermediate cases is facilitated by introducing the concept of electronegativity.

Electronegativity (*EN*) is a term that describes the relative ability of an atom to attract electrons to itself in a chemical bond. For example, in a sodium chloride molecule the chlorine atom has a large electronegativity and the sodium atom has a small electronegativity. The result is a virtually complete transfer of one electron from sodium to chlorine in the molecule. The American scientist Robert S. Mulliken proposed that electronegativity be defined as proportional to the sum of the ionization energy and the electron affinity of an atom:

$$EN = c(IE + EA)$$

Ionization energy is a measure of the ability of an atom to hold one electron, and electron affinity is a measure of the ability of an atom to attract an electron. It follows that an atom such as chlorine, which has both a large ionization energy and a large electron affinity, will have a large electronegativity. A quantitative Mulliken scale of atomic electronegativities can be obtained by assigning one atom a specific *EN* value, thus fixing the constant of proportionality, *c*. Unfortunately, not many atomic electron affinities are known accurately, thus only a few *EN* values can be calculated in this way.

A more widely applied quantitative treatment of electronegativity was introduced by the American chemist Linus Pauling in the early 1930's. The Pauling electronegativity value for a specific atom is obtained by comparing the bond energies of certain molecules containing that atom. If the bonding electrons were shared equally in a molecule *AB*, it would be reasonable to assume that the bond energy of *AB* would be the geometric mean of the bond energies of the molecules A_2 and B_2. However, the bond energy of an *AB* molecule almost always is greater than the geometric mean of the bond energies of A_2 and B_2. An example that illustrates this is the HF molecule. The bond energy of HF is 135 kcal mole^{-1}, whereas the bond energies of H_2 and F_2 are 103 kcal mole^{-1} and 33 kcal mole^{-1}, respectively. The geometric mean of the latter two values is $(33 \times 103)^{1/2} = 58$ kcal mole^{-1}, which is much less than the observed bond energy of HF. This "extra" bond energy (designated Δ) in an *AB* molecule is assumed to be a consequence of the partial ionic character of the bond due to electronegativity differences between atoms *A* and *B*. In this model the electronegativity difference between two atoms *A* and *B* is defined as

$$EN_A - EN_B = 0.208\Delta^{1/2} \qquad (2\text{–}1)$$

in which EN_A and EN_B are the electronegativities of atoms *A* and *B*, and Δ is the extra bond energy in kilocalories per mole. The extra bond energy is calculated from the equation

$$\Delta = DE_{AB} - [(DE_{A_2})(DE_{B_2})]^{1/2}$$

in which DE is the particular bond dissociation energy.

In Equation 2–1 the factor 0.208 converts kilocalories per mole to electron volts. The square root of Δ is used because it gives a more consistent set of atomic electronegativity values. Since only differences are obtained from Equation 2–1, one atomic electronegativity must be assigned a specific value, then the other values can be calculated easily. In a widely adopted version of the Pauling scale the most electronegative atom, fluorine, is assigned an electronegativity of 3.98. A compilation of EN values based on this scale is given in Table 2–4.

Table 2–4. Atomic electronegativities[a]

I	II	III	II	II	II	II	II	II	II	I	II	III	IV	III	II	I
H																
2.20																
Li	Be											B	C	N	O	F
0.98	1.57											2.04	2.55	3.04	3.44	3.98
Na	Mg											Al	Si	P	S	Cl
0.93	1.31											1.61	1.90	2.19	2.58	3.16
K	Ca	Sc	Ti	V	Cr	Mn	Fe	Co	Ni	Cu	Zn	Ga	Ge	As	Se	Br
0.82	1.00	1.36	1.54	1.63	1.66	1.55	1.83	1.88	1.91	1.90	1.65	1.81	2.01	2.18	2.55	2.96
Rb	Sr	Y	Zr		Mo			Rh	Pd	Ag	Cd	In	Sn	Sb		I
0.82	0.95	1.22	1.33		2.16			2.28	2.20	1.93	1.69	1.78	1.96	2.05		2.66
Cs	Ba	La			W			Ir	Pt	Au	Hg	Tl	Pb	Bi		
0.79	0.89	1.10			2.36			2.20	2.28	2.54	2.00	2.04	2.33	2.02		
		Ce	Pr	Nd		Sm		Gd		Dy	Ho	Er	Tm		Lu	
		1.12	1.13	1.14		1.17		1.20		1.22	1.23	1.24	1.25		1.27	
			(III)	(III)		(III)		(III)		(III)	(III)		(III)		(III)	
					U	Np	Pu									
					1.38	1.36	1.28									
					(III)	(III)	(III)									

[a] Roman numerals refer to the oxidation numbers of the atoms in the molecules used in the calculation of atomic electronegativity values.

Electronegativity is a useful concept for qualitatively describing the sharing of electrons in a bond between two atoms of different elements. In the case of sodium and chlorine the difference in the electronegativities of the two atoms is so large that there effectively is complete transfer of the electron pair. However, there are many molecules in which the bond between dissimilar atoms is described better as covalent with some ionic character.

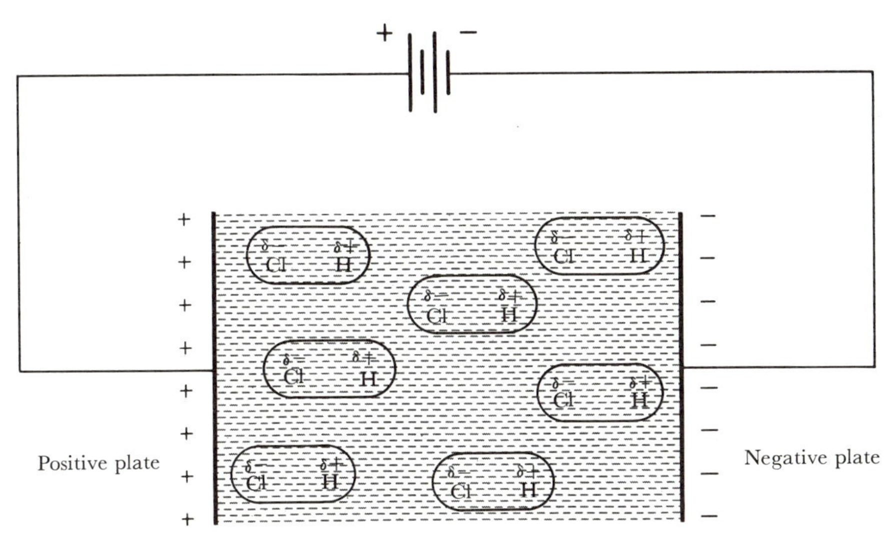

2–6
Effect of an electric field on the alignment of polar molecules. A polar molecule tends to align in an electric field, thereby maximizing the electrostatic attraction between the plates. In this illustration the preferred alignment of the polar molecule HCl is shown.

2–10 A COVALENT BOND WITH IONIC CHARACTER; THE HCl MOLECULE

The bond in a molecule composed of a hydrogen atom and a chlorine atom, HCl, is neither purely covalent nor purely ionic. In the Lewis structure for HCl an electron-pair bond between H· and ·$\ddot{\underset{..}{\text{Cl}}}$: is formed, thereby associating two electrons with hydrogen and eight electrons with chlorine (H—$\ddot{\underset{..}{\text{Cl}}}$:). The covalent structure is a more accurate representation of the bonding in HCl than an ion pair, such as we formulated for sodium chloride, because the electronegativity of the hydrogen atom is much larger (and closer to that of chlorine) than the electronegativity of the sodium atom. Although complete transfer of the pair of electrons to the chlorine atom in H—$\ddot{\underset{..}{\text{Cl}}}$: does not occur, the electron density in the bond is more concentrated in the region of the chlorine atom than in the region of the hydrogen atom. This unequal charge distribution is illustrated by

$$\overset{\delta+}{\text{H}}\text{—}\overset{\delta-}{\ddot{\underset{..}{\text{Cl}}}}\text{:}$$

There is a small, net positive charge associated with the hydrogen atom because the electron pair is "pulled" toward the chlorine atom, which acquires a small net negative charge. Thus the HCl bond is said to have ionic character.

A molecule such as HCl is polar. The measure of the tendency of a polar molecule to become aligned in an electric field (Figure 2–6) gives a quantity that is known as the electric *dipole moment*, which is related to the net charge separation in the most stable electronic state of the molecule. An HCl molecule has an electric dipole moment due to unequal sharing of the two electrons in the bond. An H_2 molecule (covalent bond) has a zero dipole moment, whereas an NaCl molecule (ionic bond) has a very large dipole moment. We will present a more detailed discussion of dipole moments and the properties of polar molecules in subsequent chapters.

2–11 LEWIS STRUCTURES FOR POLYATOMIC MOLECULES

Communication among chemists commonly involves a language filled with Lewis structures for the molecules under discussion. It is important to develop considerable facility in formulating these "line and dot" structures for all types of molecules. In the following sections we will discuss several representative polyatomic molecules.

Carbon tetrachloride molecule

In CCl_4 the four chlorine atoms are bonded to the carbon atom. We could begin writing the Lewis structure for CCl_4 by making electron-pair bonds with the unpaired electrons in the ground-state carbon atom:

```
      Cl:
     /
  :C
     \
      Cl:
```

But in $:CCl_2$ the bonding capacity of carbon is not saturated. That is, the carbon atom can accommodate eight valence electrons, but in $:CCl_2$ it has only six electrons. If energy is added to the carbon atom to unpair its two 2*s* electrons, four bonds can be made between the carbon atom and the chlorine atoms. Thus the correct Lewis structure for carbon tetrachloride is

```
       :Cl:
        |
  :Cl—C—Cl:
        |
       :Cl:
```

The carbon atom now is saturated; that is, it has achieved a closed-shell configuration. The energy required to unpair the 2*s* electrons is small compared to the energy released in forming the two additional single bonds. Therefore carbon tetrachloride is much more stable than the system $CCl_2 + 2Cl$. Notice that there are eight valence electrons associated with each atom. The valence electrons in CCl_4 not involved in bonding are the unshared-pair (or lone-pair) electrons associated with the chlorine atoms.

Carbon tetrachloride is a covalent molecule. Although carbon and chlorine both are nonmetals with relatively large electronegativities, atomic chlorine has a larger electronegativity than atomic carbon, but not large enough to effect a complete transfer of the pair of electrons in each bond to give the ionic structure

$$\begin{array}{c} Cl^- \\ Cl^- C^{4+} Cl^- \\ Cl^- \end{array}$$

However, in each bond there is partial ionic character of the type $C^{\delta+}Cl^{\delta-}$ because each pair of electrons is associated more with the chlorine atom than with the carbon atom. The symbol $Cl^{\delta-}$ is used to indicate a "partial negative charge," which is less than that if there were complete electron transfer to the chlorine atom. Similarly, $C^{\delta+}$ implies a corresponding partial positive charge associated with the carbon atom. However, because carbon tetrachloride is symmetrical the molecule has no *net* dipole moment. The individual bond dipoles, $C\underline{^{\delta+}}Cl^{\delta-}$, are oriented in different directions and the shape of the molecule is such that they cancel each other. Thus a zero dipole moment does not always mean that there is equal sharing of electron pairs in bonds. A zero dipole moment may arise from the cancellation of several bond dipoles due to the shape of the molecule. We will discuss this topic in more detail in Chapter 4.

Ammonia molecule

The correct Lewis structure for ammonia, NH_3, is

$$\begin{array}{c} H\text{—}\ddot{N}\text{—}H \\ | \\ H \end{array}$$

with three N–H bonds and one unshared pair of electrons. There are eight valence electrons around the nitrogen atom and two electrons associated with each hydrogen atom, thereby providing a closed-shell configuration for each atom.

Magnesium chloride, an ionic molecule

A magnesium atom has a small electronegativity and atomic chlorine has a large electronegativity, thus magnesium chloride requires a structure showing an ionic bond. One electron is transferred from the magnesium atom to each of the chlorine atoms in $MgCl_2$. The correct structure indicates the charges:

$$[:\ddot{\underset{..}{Cl}}:^-][Mg^{2+}][:\ddot{\underset{..}{Cl}}:^-]$$

There are two large bond dipoles ($\overset{\leftarrow\!+}{\underset{-}{Cl}}\ \overset{++}{Mg}\ \overset{+\!\rightarrow}{\underset{-}{Cl}}$) in opposite directions and of equal magnitude (since the chloride ions are equivalent). These two bond dipoles cancel because the geometrical structure of the molecule is linear, as indicated in the preceding structure. This is another case of a zero dipole moment resulting from a symmetrical molecular shape.

Ammonium chloride molecule

The ammonium chloride molecule, NH_4Cl, contains NH_4^+ and Cl^- ions. First we will describe the bonding in the ammonium ion, NH_4^+. The Lewis structure for the nitrogen atom is :Ṅ̤· and for each hydrogen atom it is H·. However, the NH_4^+ ion has one positive charge, which means that one of the nine electrons has been lost. Since all the hydrogen atoms in the ion are equivalent we give the nitrogen atom the positive charge, ·Ṅ̤·$^+$, and write

```
     H
     |
H—N⁺—H
     |
     H
```

Thus the correct Lewis structure of the NH_4^+ ion has four single bonds and no unshared electrons.

The charge on the nitrogen atom is called a *formal charge*. Formal charges are assigned in the following way. In a bond involving two atoms joined by an electron pair, we consider that each atom "owns" one of the electrons. Thus each hydrogen atom in the ammonium ion owns one electron. But each neutral hydrogen atom has one electron to begin with, so it has no formal charge in the molecule. However, the nitrogen atom has only four electrons in the NH_4^+ ion. Since this is one less than the five electrons in atomic nitrogen, we assign a formal charge of +1 to the nitrogen atom in NH_4^+.

Finally, we assign the electron that was removed from NH_4 (to give NH_4^+) to the chlorine atom to give the "ion-pair" structure for ammonium chloride molecule:

```
 ┌     H     ┐
 |     |     |
 | H—N⁺—H   |[:C̤̈l:⁻]
 |     |     |
 └     H     ┘
```

In summary, the N–H bonds in NH_4^+ are viewed as covalent with some ionic character, whereas the NH_4^+ ion is attached to Cl^- by an ionic bond.

2–12 MOLECULES WITH DOUBLE AND TRIPLE BONDS

Now we consider molecules in which more than one electron pair are involved in a bond between two atoms. An example is the ethylene molecule, C_2H_4, in which the four hydrogen atoms are attached to the two carbon atoms and the two carbon atoms are attached to each other. First we can make the bonds to the hydrogen atoms:

```
H           H
  \        /
   C:   :C
  /        \
H           H
```

Then by forming a single bond between the carbon atoms and substituting a line for the bonding electron pair we have

```
H       H
 \     /
  Ċ—Ċ
 /     \
H       H
```

One unshared valence electron remains on each carbon atom. Counting the electrons we find that there are seven valence electrons associated with each carbon atom and two electrons with each hydrogen atom. If we make an additional electron-pair bond between the two carbon atoms, then each carbon atom has the required eight electrons:

```
H       H
 \     /
  C=C
 /     \
H       H
```

The bond between the two carbon atoms involves two electron pairs and is called a *double bond*. Therefore the carbon–carbon bond order in ethylene is two.

There are many compounds that require Lewis structures with double or triple bonds.

Structure	Bond
>C=C<	carbon–carbon double bond
—C≡C—	carbon–carbon triple bond
—N̈=N̈—	nitrogen–nitrogen double bond
:Ö=Ö:	oxygen–oxygen double bond
:N≡N:	nitrogen–nitrogen triple bond
>C=N̈—	carbon–nitrogen double bond
—C≡N:	carbon–nitrogen triple bond
>C=Ö:	carbon–oxygen double bond

The preceding structures include that for the nitrogen molecule, N_2, which has a bond order of three. Nitrogen contributes about 80% of the earth's atmosphere, and it generally is regarded as being almost entirely unreactive. The inert character of nitrogen results because atoms of the element are very strongly bound together in diatomic molecules.

An interesting problem arises in the Lewis structural formulation of the common air pollutant molecule, nitric oxide (NO). A closed-shell configuration cannot be constructed for NO because there is an odd number of valence electrons. Nitric oxide has eleven valence electrons, five valence electrons originally associated with the nitrogen atom and six electrons with the oxygen atom. Thus either N or O will own only seven electrons in the NO molecule. We choose N because it is less electronegative than O. Therefore the best structure for NO is

$$:\dot{N}{=}\underset{..}{O}:$$

Lewis structures for molecules such as NO, which have an odd number of electrons, necessarily cannot have closed shells associated with each atom. At least one atom, nitrogen in the NO example, is left with an "open shell."

The odd electron in NO is unpaired. Consequently we would predict nitric oxide to be paramagnetic, a prediction that is in agreement with experimental data. A multiple-bonded molecule with particularly vexing magnetic properties (for Lewis structural theory) is O_2, which is known to have two unpaired electrons in its ground state and to be paramagnetic. An unusual structure such as $:\dot{O}{\equiv}\dot{O}:$ or $:\underset{.}{\ddot{O}}{-}\underset{.}{\ddot{O}}:$ would be required to explain this magnetic behavior. However, the observed bond length and bond energy of O_2 are completely consistent with the simple double-bond structure $:\ddot{O}{=}\ddot{O}:$. We will see in the next chapter that the molecular orbital theory provides a satisfactory explanation of both the paramagnetism and the bond properties of the oxygen molecule.

2–13 BONDING TO HEAVIER ATOMS

The octet rule has been extremely valuable as a guide in writing electronic formulas. For second-row nonmetallic elements (B, C, N, O, F) exceptions to the rule are very rare. It is easy to rationalize why this is so. Atoms of the second-row elements have stable 2*s* and 2*p* orbitals, and the "magic number" of eight corresponds to the closed valence-orbital configuration $2s^2 2p^6$. Adding a ninth, tenth, or larger number of electrons to such a configuration is impossible because the next atomic orbital available to a second-row element is the highly energetic 3*s* orbital.

Beyond the second row in the periodic table the octet rule is not obeyed with such satisfying regularity. However, it remains a useful rule, as illustrated by molecules such as PH_3, PF_3, H_2S, and SF_2:

H—P̤̈—H | H	:F̤̈—P̤̈—F̤̈: | :F̤̈:	H—S̤̈: | H	:F̤̈—S̤̈: | :F̤̈:
phosphine	phosphorus trifluoride	hydrogen sulfide	sulfur difluoride

Atoms of the heavier elements do more than obey the octet rule. Some of them show a surprising ability to bind more atoms (or associate with more electron pairs) than would be predicted from the octet rule. For example, phosphorus and sulfur form the compounds PF_5 and SF_6, respectively. Lewis structures for these compounds use all the valence electrons of the heavy element in bonding:

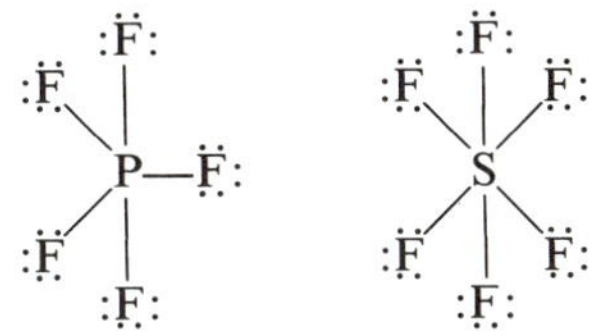

```
      :F:              :F:
  :F   |           :F   |   F:
     \ |              \ | /
      P—F:              S
     / |              / | \
  :F   |           :F   |   F:
      :F:              :F:
```

phosphorus pentafluoride   sulfur hexafluoride

That phosphorus shares ten electrons and sulfur shares twelve electrons obviously violates the octet rule. The theory of atomic structure helps us see why the violation has occurred. The noble gas in the third row with phosphorus and sulfur is argon. The argon electronic structure fills the 3*s* and 3*p* orbitals, but leaves the fivc 3*d* orbitals vacant. If some of these 3*d* orbitals are used for electron-pair sharing, extra bonds are possible. The atomic theory thus provides an explanation of the enhanced bonding versatility of elements in the third row and beyond.

Perhaps the most important consequence of the use of *d* orbitals is the existence of an important series of oxyacids. The most well-known examples are phosphoric acid (H_3PO_4), sulfuric acid (H_2SO_4), and perchloric acid ($HClO_4$). It is possible to write a Lewis structure for sulfuric acid that obeys the octet rule,

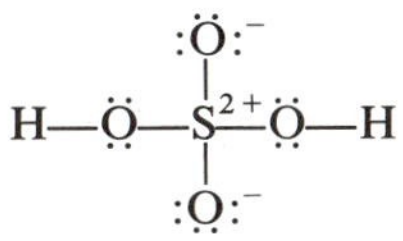

```
         :O:⁻
          |
  H—O—S²⁺—O—H
          |
         :O:⁻
```

However, examination of this structure reveals that a formal charge of +2 is on the sulfur atom. Development of a large positive formal charge on an electronegative nonmetal atom is not very reasonable. The formal charge can be removed if we write two S–O double bonds, thereby allowing the sulfur atom to share 12 electrons:

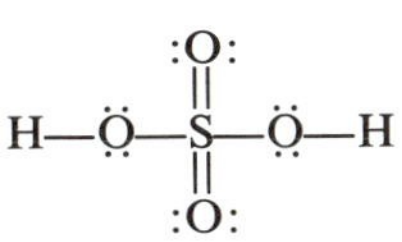

```
         :O:
          ||
  H—O—S—O—H
          ||
         :O:
```

sulfuric acid

Similar Lewis formulas can be written for other oxyacids:

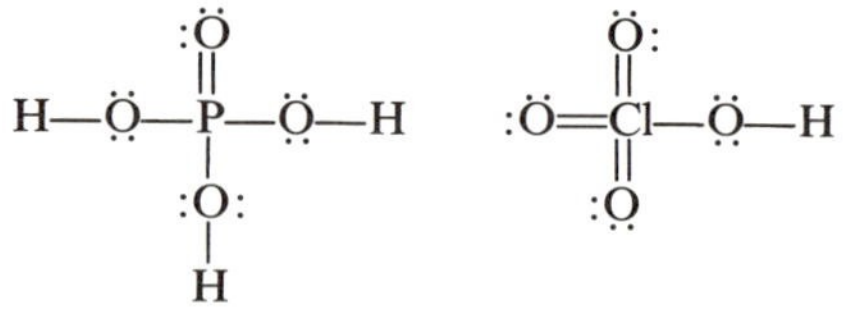

```
        :Ö                    Ö:
         ||                   ||
H—Ö—P—Ö—H          :Ö=Cl—Ö—H
         |                    ||
        :Ö:                  :Ö
         |
         H
```

phosphoric acid　　　　perchloric acid

2–14 RESONANCE

There are molecules and ions for which more than one satisfactory Lewis formula can be drawn. For example, the nitrite ion, NO_2^-, can be formulated as either

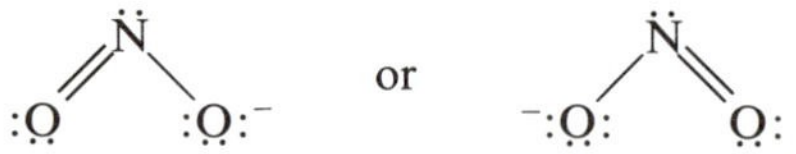

```
      N̈                        N̈
    //  \          or        /  \\
  :Ö     :Ö:⁻             ⁻:Ö:    Ö:
```

In either case the octet rule is satisfied. If either of these structures were the "correct" one, the ion would have two distinguishable nitrogen–oxygen bonds, one single and one double. Double bonds are shorter than single bonds, but structural studies of NO_2^- show that the two N–O bonds are indistinguishable.

Consideration of NO_2^- and many other molecules and ions shows that our simple scheme for counting electrons and assigning them to the valence shells of atoms as bonds or unshared pairs is not entirely satisfactory. Fortunately, the simple model is altered fairly easily to fit many of the awkward cases. The problem with NO_2^- is that the ion is actually more symmetrical than either one of the Lewis electronic structures that we wrote. However, if we took photographs of the two formulas shown previously and superimposed the pictures, we would obtain a new formula having the same symmetry as the molecule. The photographic double-exposure method is the same as writing a formula such as

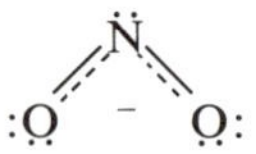

```
      N̈
    //  \\
  :Ö  ⁻  Ö:
```

This formula would imply, "NO_2^- is a symmetrical ion, having partial double-bond character in each of the N–O bonds." For some purposes the formula is adequately informative. However, keeping track of the electrons in such a formula requires the addition of some rather special notation. What we actually do most of the time in such situations is to write two or more Lewis formulas and connect them with a symbol that means: "Superimpose these formulas to get a reasonable representation of the molecule." Applied to NO_2^- the formulas are

The double-headed arrow is the symbol reserved for this purpose. It should not be confused with the symbol consisting of two arrows pointing in opposite directions, ⇆, which indicates that a reversible chemical reaction occurs. The double-headed arrow conveys no implication of dynamic action.

The method of combining two or more structural formulas to represent a single chemical species is called the *resonance method*. The method is used not only for the construction of electronic formulas, but also as the basis of one method for doing approximate quantum-mechanical analyses of molecular structures.

When we consider the benzene molecule, C_6H_6, which has six carbon atoms arranged in a ring, we can draw two formulas that are equally satisfactory:

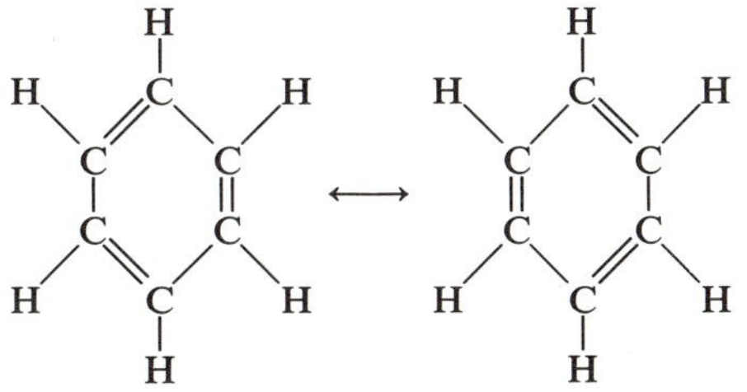

Both resonance structures show the ring to be composed of alternate single and double bonds. However, structural studies reveal that all of the carbon–carbon bond distances are equal. The full symmetry of the molecule is indicated by a double-headed arrow between the two structures.

Resonance notation is required in many cases other than those in which it is demanded by symmetry. For example, compare two well-known anions, nitrate (NO_3^-) and nitroamide (^-O_2NNH). Nitrate has threefold symmetry, so we can write a set of three equivalent resonance structures:

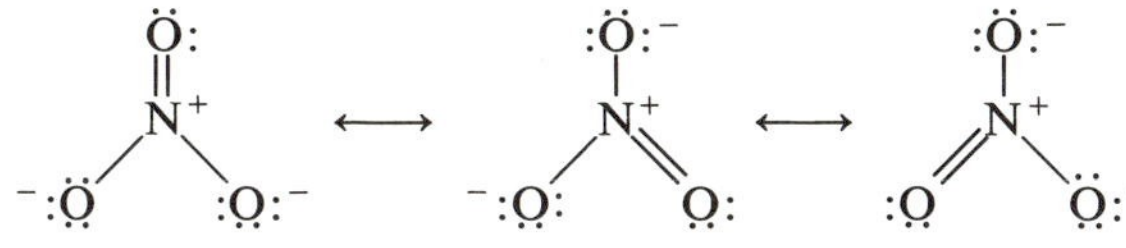

For the nitroamide ion we can write two equivalent structures, plus a third

that is not equivalent to the other two:

$$
\underset{\text{I}}{{}^{-}\!:\ddot{\text{N}}\text{—H},\ \text{N}^{+},\ {}^{-}\!:\ddot{\text{O}}\text{—},\ \text{=}\ddot{\text{O}}\!:}
\longleftrightarrow
\underset{\text{II}}{{}^{-}\!:\ddot{\text{N}}\text{—H},\ \text{N}^{+},\ :\ddot{\text{O}}\text{=},\ \text{—}\ddot{\text{O}}\!:^{-}}
\longleftrightarrow
\underset{\text{III}}{\ddot{\text{N}}\text{—H},\ \text{=N}^{+},\ {}^{-}\!:\ddot{\text{O}}\text{—},\ \text{—}\ddot{\text{O}}\!:^{-}}
$$

Common sense tells us that all three structures should contribute to our description of the ion. Since the structures are not equivalent, the resonance symbol no longer means: "Mix these structures equally in your thinking." It merely means: "Mix them." Therefore no quantitative implications are intended by the double-headed arrow. When we become semiquantitative in our description we state that Structure III "contributes" more to the structure of the nitroamide ion than either of the equivalent Structures I and II because III places both formal negative charges on the oxygen atoms.

Finally, we will discuss the anion obtained by removing the two protons in sulfuric acid, the sulfate ion, SO_4^{2-}. As in the case of H_2SO_4 an octet-rule structure with only single bonds can be written by assigning three lone pairs to each oxygen atom:

$$
{}^{-}\!:\ddot{\text{O}}\text{—}\overset{\overset{:\ddot{\text{O}}:^{-}}{|}}{\underset{\underset{:\ddot{\text{O}}:^{-}}{|}}{\text{S}^{2+}}}\text{—}\ddot{\text{O}}\!:^{-}
$$

However, if we consider the large positive formal charge on the sulfur atom we conclude that this is not a particularly appropriate structure. A much better representation of the bonding in SO_4^{2-} removes the +2 formal charge on the central sulfur atom by forming two sulfur–oxygen double bonds.

There are six equivalent structures with two S=O bonds and two S–O bonds. Thus we represent the bonding in SO_4^{2-} as a resonance hybrid of the following six equivalent structures:

$$
\underset{\text{I}}{:\ddot{\text{O}}\text{=}\overset{\overset{:\ddot{\text{O}}}{\|}}{\underset{\underset{:\ddot{\text{O}}:^{-}}{|}}{\text{S}}}\text{—}\ddot{\text{O}}\!:^{-}}
\longleftrightarrow
\underset{\text{II}}{:\ddot{\text{O}}\text{=}\overset{\overset{:\ddot{\text{O}}:^{-}}{|}}{\underset{\underset{:\ddot{\text{O}}:^{-}}{|}}{\text{S}}}\text{=}\ddot{\text{O}}\!:}
\longleftrightarrow
\underset{\text{III}}{:\ddot{\text{O}}\text{=}\overset{\overset{:\ddot{\text{O}}:^{-}}{|}}{\underset{\underset{:\ddot{\text{O}}}{\|}}{\text{S}}}\text{—}\ddot{\text{O}}\!:^{-}}
$$

$$
\longleftrightarrow
\underset{\text{IV}}{{}^{-}\!:\ddot{\text{O}}\text{—}\overset{\overset{:\ddot{\text{O}}}{\|}}{\underset{\underset{:\ddot{\text{O}}:^{-}}{|}}{\text{S}}}\text{=}\ddot{\text{O}}\!:}
\longleftrightarrow
\underset{\text{V}}{{}^{-}\!:\ddot{\text{O}}\text{—}\overset{\overset{:\ddot{\text{O}}}{\|}}{\underset{\underset{:\ddot{\text{O}}}{\|}}{\text{S}}}\text{—}\ddot{\text{O}}\!:^{-}}
\longleftrightarrow
\underset{\text{VI}}{{}^{-}\!:\ddot{\text{O}}\text{—}\overset{\overset{:\ddot{\text{O}}:^{-}}{|}}{\underset{\underset{:\ddot{\text{O}}}{\|}}{\text{S}}}\text{=}\ddot{\text{O}}\!:}
$$

The resonance hybrid of the six equivalent structures (I–VI) of SO_4^{2-} would have an average S–O bond order of 1 1/2. In accord with this model of partial double-bond character is the fact that the observed S–O bond length in SO_4^{2-} (1.49 Å) is 0.21 Å shorter than the standard S–O single-bond length of 1.70 Å, which is obtained by adding the atomic radii of sulfur (1.04 Å) and oxygen (0.66 Å) (see Figure 2–1).

SUGGESTIONS FOR FURTHER READING

A. Companion, *Chemical Bonding*, McGraw-Hill, New York, 1964.

H. B. Gray, *Electrons and Chemical Bonding*, Benjamin, Menlo Park, Calif., 1965.

J. L. Hall and D. A. Keyworth, *Brief Chemistry of the Elements*, Benjamin, Menlo Park, Calif., 1971.

L. Pauling, *The Chemical Bond*, Cornell Univ. Press, Ithaca, N.Y., 1967.

R. L. Rich, *Periodic Correlations*, Benjamin, Menlo Park, Calif., 1965.

G. E. Ryschkewitsch, *Chemical Bonding and the Geometry of Molecules*, Reinhold, New York, 1962.

R. T. Sanderson, *Chemical Periodicity*, Reinhold, New York, 1960.

QUESTIONS AND PROBLEMS

1. Write orbital electronic configurations and then draw Lewis structures for atomic sodium, silicon, phosphorus, and sulfur. How many unpaired electrons are there in each atom?
2. The ionization energies of francium (Fr) and astatine (At) are not given in Table 2–1 because these elements are not available in large quantities and no accurate *IE* measurements have been made. Using the information available in Table 2–1 estimate the values of *IE* for francium and astatine atoms.
3. Explain why the electron affinities of both silicon and sulfur are larger than that of phosphorus.
4. Make a plot of IE_2 values versus atomic number for the elements helium through calcium. Explain the differences between this plot and that given for the IE_1 values (Figure 2–3). Why are the maximum IE_2 values not those of the noble gases?
5. The ionization energy of atomic hydrogen is 13.6 eV and the electron affinity is 0.756 eV. Convert these quantities to cm^{-1} and to kcal $mole^{-1}$.
6. Predict the relative effective radii of the species H^-, He, and Li^+. Explain your choice.
7. Xenon forms a number of interesting molecules and ions with fluorine and oxygen. Write a Lewis structure for each of the following: XeO_4, XeO_3, XeF_8^{2-}, XeF_6, XeF_4, XeF_2, and XeF^+. Show the placement of formal charges in the Lewis structures. Avoid structures with formal charge separation, if possible. Based on expected trends in effective atomic radii, predict whether the Xe–F bond length in XeF_4 will be longer, or shorter, than the I–F bond length in the related ion IF_4^-.

8. Show that e^2 has a value of 332 kcal mole^{-1} Å.
9. The bond distance in diatomic LiF is 1.52 Å. Assuming ionic bonding calculate the energy required to dissociate LiF into Li^+ and F^-.
10. Calculate the electronegativity of hydrogen (EN_H) assuming the value 3.98 for EN_F from Table 2–4. Do this calculation for both the Pauling and the Mulliken scales. The value you obtain on the Pauling scale will not agree exactly with that given in Table 2–4 because many *EN* differences (Equation 2–1) were averaged to give the best values reported in the table.
11. Nitrogen forms a trifluoride, NF_3, but NF_5 does not exist. For phosphorus both PF_3 and PF_5 are known. Write Lewis structures for NF_3, PF_3, and PF_5. Discuss possible explanations for the fact that PF_5 is stable, whereas NF_5 is not. From your treatment, which of the following molecules would you expect not to exist: OF_2, OF_4, OF_6, SF_2, SF_4, and SF_6? Write Lewis structures and appropriate comments to support your case.
12. Write Lewis structures for CO_2 and SO_2. Are the C–O bonds primarily ionic or covalent? The SO_2 molecule has a dipole moment, whereas CO_2 does not. What shape do you expect for each molecule?
13. The acetylene molecule, HCCH, is linear. Write a Lewis structure for acetylene. Do you expect the C–C bond to be longer in C_2H_2 than in C_2H_4? Compare the energies of the C–C bonds in C_2H_4 and C_2H_2. Is C_2H_2 polar or nonpolar?
14. Iodine forms several oxyions of the type IO_x^{n-}. Write Lewis structures for IO_3^-, IO_4^-, and IO_6^{5-}. Predict the relative I–O bond lengths in these oxyions.
15. Write a Lewis structure for S_2. Do you expect the molecule to be paramagnetic or diamagnetic?
16. Write Lewis structures for BF_3 and NO_3^-. Do these molecular species have anything in common? The dipole moment of BF_3 is zero. What geometrical structure do you expect for BF_3? What geometrical structure might NO_3^- have?
17. Write Lewis structures for CN^- and CO. Is the C–O bond length in CO shorter than in CO_2? Explain.
18. Assuming an ionic model calculate standard bond dissociation energies for the following molecules: CsF, KBr, and LiF. Bond lengths and experimental bond-energy data (for comparison) are given in Table 3–5.
19. Write Lewis structures for H_2O and HF. The water molecule is nonlinear. Do H_2O and HF have dipole moments?
20. Write Lewis structures for the following molecules and ions. Show resonance structures if appropriate. Also indicate in each case if there are formal charges on one or more atoms.

a) FBr b) S_2 c) Cl_2 d) P_2
e) NCO^- f) CNO^- g) $BeCl_2$ h) CS_2
i) BF_3 j) SO_3 k) CO_3^{2-} l) CF_4
m) $SiBr_4$ n) BF_4^- o) NCl_3 p) PF_3

q) CH_3^- r) SF_2 s) XeO_3 t) SO_2

u) SF_6 v) Na_2O w) ClO_2

x) N_2F_2 y) CsF z) SrO

21. Calculate the standard bond energies of RbF, CsBr, NaI, and KCl. Assume an ionic model. Bond-length data and experimental bond energies are given in Table 3–5.

22. Calculate the dissociation energies of BeO and CaO to M^{2+} and O^{2-} ions using an ionic bonding model. Bond distances are given in Table 3–5.

23. Write the electronic configuration ($1s^2 2s^2 \cdots$) for the following: F^-, Na^+, Ne, O^{2-}, and N^{3-}. What would you predict about the relative sizes of these species?

24. Which atom in each of the following pairs would you expect to have the larger electron affinity (*EA*)? (a) Cu or Zn; (b) K or Ca; (c) S or Cl; (d) H or Li; (e) As or Ge.

25. The electronic structure of the thiocyanate ion, NCS^-, can be represented as a hybrid of two resonance structures. Write these two structures and give the C–N and C–S bond orders for each structure.

26. For each of the following cases give the Lewis structure of a known chemical example:
    a) a diatomic molecule with one unpaired electron
    b) a triatomic molecule with two double bonds
    c) a diatomic molecule with formal charge separation
    d) a diatomic molecule with partial ionic character
    e) an alkaline-earth oxide
    f) a molecule or ion with two equivalent resonance structures
    g) a molecule or ion with three equivalent resonance structures

27. The atoms of the yet-to-be-discovered "hypotransition" elements, starting at $Z = 121$, will have electrons in the $5g$ orbitals.
    a) How many elements will there be in the hypotransition metal series?
    b) How many of the atoms will be diamagnetic?
    c) Which electronic configurations in the series will have seven unpaired electrons?
    d) What is the maximum number of unpaired electrons an atom can have in the series? Will this be a new record for atoms in the periodic table?
    e) What is the *IE* of an electron in a $5g$ orbital of atomic hydrogen? Is this likely to be larger, or smaller, than the *IE* of a $5g$ electron in one of the hypotransition elements? Briefly explain your choice.
    f) In atomic hydrogen the $5s$, $5p$, $5d$, $5f$, and $5g$ orbitals all have the same energy. Will this be true for the hypotransition elements? If not, what will the energy order be? Explain briefly.

3

Diatomic Molecules

Although they provide a simple and convenient representation of chemical bonding in molecules, the lines for electron-pair bonds in Lewis structures tell us nothing about many interesting details of molecular and electronic structure. A more comprehensive analysis of electronic structural properties is made possible by considering the valence orbitals involved in chemical bonds. In this chapter we will introduce a powerful method of analysis known as *molecular orbital (MO) theory.*

3–1 MOLECULES WITH 1*s* VALENCE ATOMIC ORBITALS

When two hydrogen atoms are separated by relatively large distances (10 Å or more), neither atomic electron cloud is influenced significantly by the presence of the other atom [Figure 3–1(a)]. However, as the two atoms approach each other the orbitals overlap and electron density increases between the nuclei, as shown in Figure 3–1(b). At the equilibrium internuclear separation in H_2 the stable molecular orbital has a distribution of electron density throughout the two atoms, but the distribution is concentrated between the nuclei [Figure 3–1(c)]. A molecular orbital that concentrates electron density in the region between the nuclei, thereby lowering the energy of the system, is called a *bonding molecular orbital.*

It is reasonable to assume that an electron in a molecular orbital is described by the appropriate atomic wave function when it is near one particular nucleus and consequently largely influenced by that nucleus. Thus the simplest mathematical function that approximately describes the molecular orbital illustrated in Figure 3–1(c) is $1s_a + 1s_b$; that is, we add the two 1*s* mathematical functions. The square of the $1s_a + 1s_b$ function is proportional to the probability of finding an electron at a given point. If the square of the assumed molecular wave function is evaluated at all points in space, a reasonable approximation of the electron-density picture shown in Figure 3–1(c) is obtained. The graphical representation of $(1s_a + 1s_b)^2$ as a function of internuclear distance is shown in Figure 3–1(d).

The spatial-boundary picture of a molecular orbital, like that of an atomic orbital, outlines the volume that encloses most of the electron density and consequently has a shape given by the molecular wave function. In Figure 3–2 we show schematically the "formation" of the bonding molecular orbital in H_2 by "adding" two 1*s* orbitals.

Molecular orbitals are classified according to their shapes or angular properties, analogous to the classification of atomic orbitals as *s*, *p*, *d*, and so on. The molecular orbital shown in Figure 3–2 is called sigma (σ) because it is symmetrical when rotated around a line joining the nuclei. Since the orbital is also bonding, its complete shorthand designation is σ^b (read sigma b), which stands for *sigma bonding molecular orbital.*

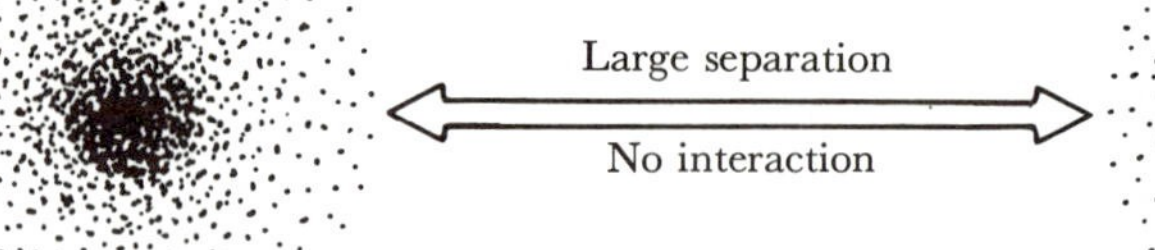
Large separation
No interaction

(a)

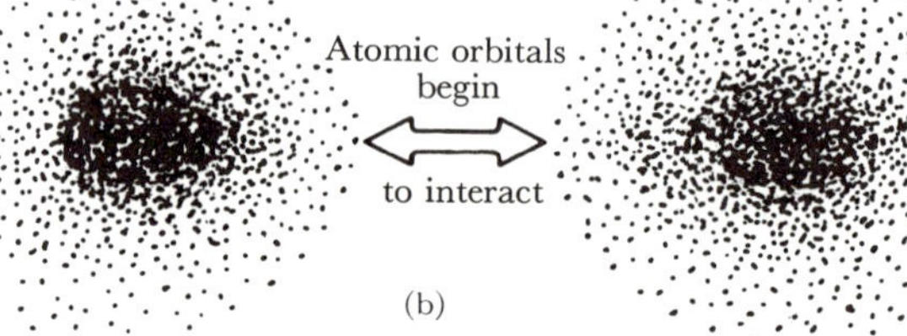
Atomic orbitals
begin
to interact

(b)

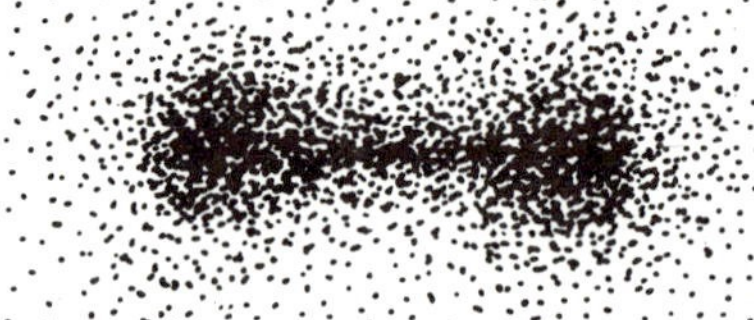

(c)

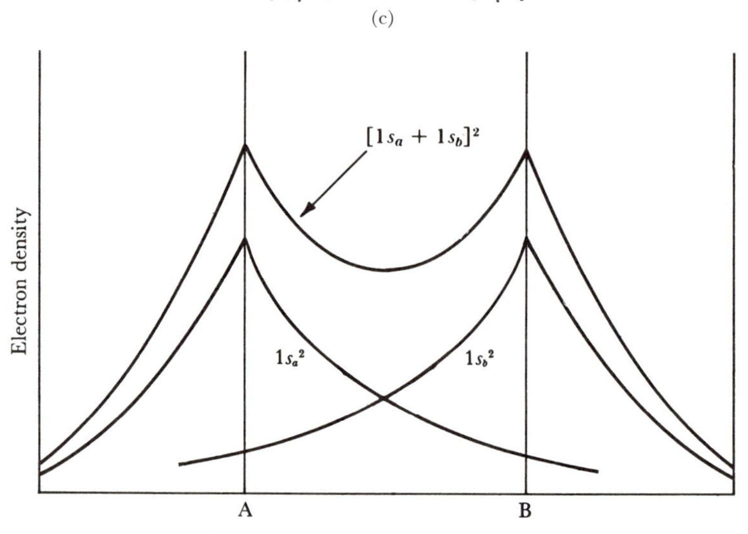
$[1s_a + 1s_b]^2$
Electron density
$1s_a{}^2$
$1s_b{}^2$
A
B
Distance along the internuclear line

(d)

◀**3–1**
Representations of atomic and molecular orbitals. (a) Electron clouds at large separation. (b) Interaction of atomic orbitals. The electron clouds become distorted and electron density increases in the region between the nuclei. (c) Molecular orbital with increased concentration of electron density between the nuclei. The nuclei are shown at their equilibrium internuclear separation. (d) Plot of estimated electron densities along the line joining the center of two hydrogen atoms. The top curve shows that the electron density between and around the two protons is greater in the molecular orbital than in the two atomic orbitals (bottom curves). It is especially important to notice that the atomic-orbital overlap produces an increased probability of finding the electrons between the two hydrogen atoms.

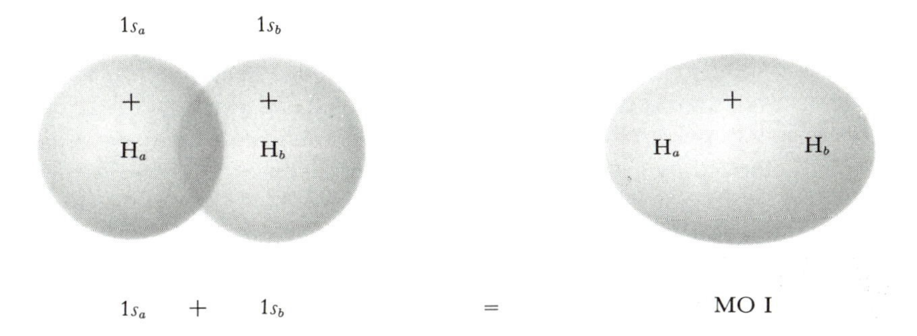

3–2
Schematic representation of the formation of the sigma bonding (σ^b) molecular orbital of H_2. In this combination of the two 1*s* valence orbitals, electron density increases in the region between the two nuclei. The σ^b molecular orbital is symmetrical around the internuclear line.

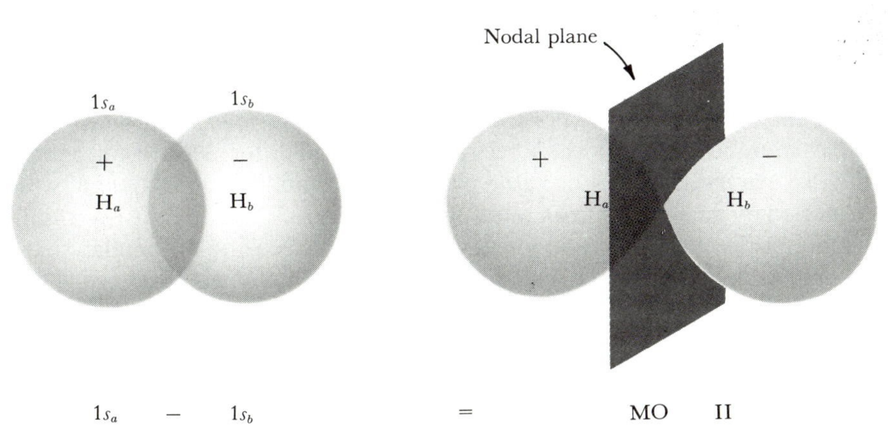

3–3
Schematic drawing of the formation of the sigma antibonding (σ^*) molecular orbital of H_2. In this subtractive combination of the two 1*s* valence orbitals, electron density decreases in the region between the nuclei. In the nodal plane there is zero probability of finding an electron. The σ^* molecular orbital is symmetrical around the internuclear line.

Suppose that there are more than two electrons to accommodate in molecular orbitals, such as there would be when two helium atoms, which contain two 1*s* electrons each, come together. Since only two of the four electrons in He_2 can be placed in the σ^b orbital, the other two electrons must occupy a more energetic orbital. This more energetic molecular orbital can be formulated as the subtractive combination of the two 1*s* valence orbitals, as shown in Figure 3–3. In the molecular orbital $1s_a - 1s_b$, electron density is reduced greatly in the overlap region and is forced away from the region between the nuclei. The electron density is zero in a plane (called a *nodal plane*) that is halfway between the nuclei and perpendicular to the internuclear line. A molecular orbital with decreased electron density between the nuclei does not "cement" the positively charged nuclei because the electron–nucleus attractions now are small relative to the nuclear repulsion. Thus this molecular orbital is less stable than the isolated atomic orbitals from which it is derived, and for this reason it is given the name "antibonding." An *antibonding molecular orbital* is characterized by greatly decreased electron density between the nuclei.

The antibonding orbital shown in Figure 3–3 also is classified as a σ molecular orbital. That is, if the orbital is rotated by any arbitrary angle around the internuclear line the orbital still looks the same. We abbreviate *sigma antibonding* as σ^* (read sigma star). Figure 3–4(a) shows the energies of σ^b and σ^* as a function of the internuclear separation.

The relative energies of the molecular orbitals commonly are given at the equilibrium internuclear separation. We have shown that from two 1*s* atomic orbitals, which have the same energy, we can construct two molecular orbitals. The bonding molecular orbital is lower in energy than the original atomic orbitals and the energy of the antibonding orbital is higher [Figure 3–4(b)].

Net bonding

The net number of electron-pair bonds in a molecular system is equal to the total number of electron pairs that can be placed in bonding orbitals minus the total number that are forced (because of the Pauli principle) into antibonding orbitals. An electron in a bonding orbital gives stability to the system, which is canceled by an electron in an antibonding orbital. We divide the net number of bonding electrons by two to preserve the idea of electron-pair bonds; that is, two net bonding electrons equals one "bond line" in a Lewis structure.

3–4▶
(a) The energy of a molecule with electrons in the bonding orbital falls to a minimum at the equilibrium internuclear separation. The energy of a molecule with electrons in the antibonding orbital is always greater than the energy of completely separated atoms; the energy of such a molecule increases steadily as the atoms are brought closer together. (b) Relative energies of the two molecular orbitals for the hydrogen molecule, and the two atomic 1*s* orbitals from which they came. The bonding MO is lower in energy, and the antibonding higher, than the energy of the original atomic orbitals.

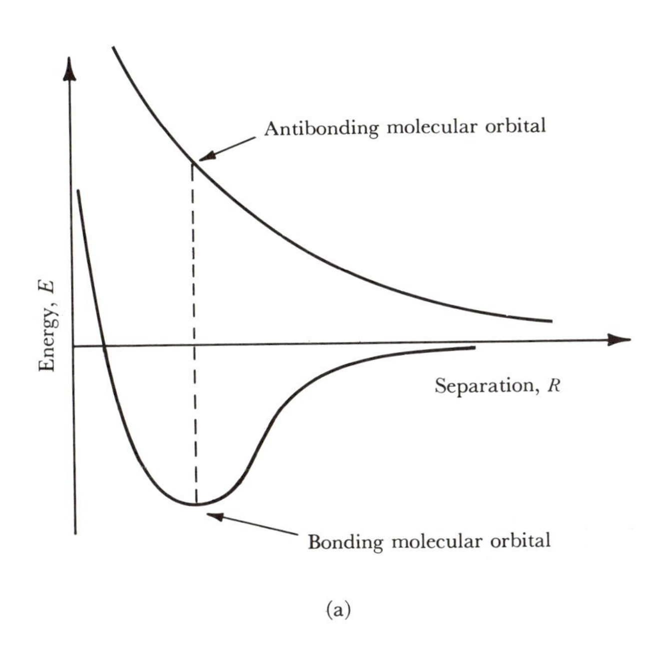
Antibonding molecular orbital
Energy, E
Separation, R
Bonding molecular orbital

(a)

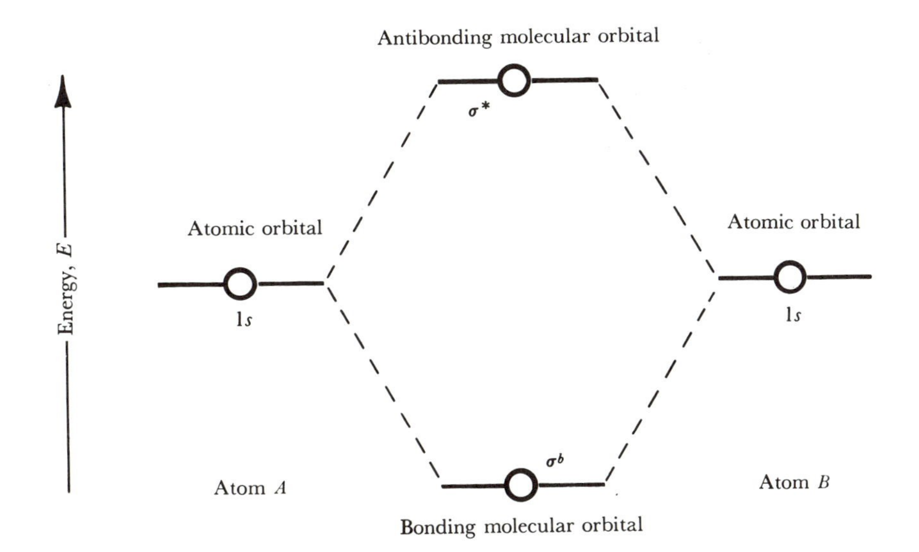
Antibonding molecular orbital
σ*
Energy, E
Atomic orbital
1s
Atomic orbital
1s
σb
Atom A
Atom B
Bonding molecular orbital

Molecule

(b)

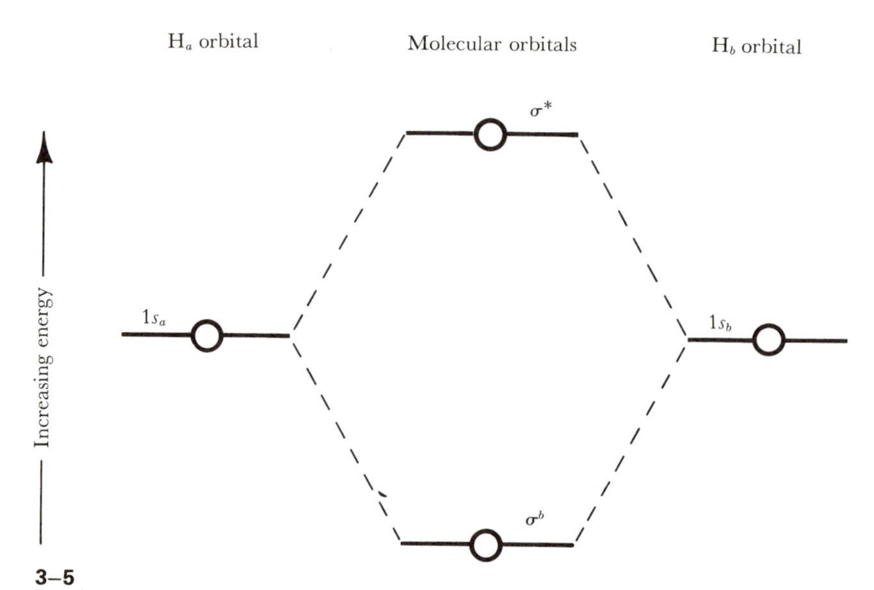

3–5
Relative molecular-orbital energies for H_2^+. This energy scheme also is appropriate for the structures of H_2, He_2^+, and He_2.

Let us consider in detail the molecular orbital theory for the simple molecules H_2^+, H_2, He_2^+, and He_2. Electronic structures are built for molecules in a way completely analogous to the *Aufbau* process for individual atoms. That is, the electronic structures of the molecules are built by putting electrons in the lowest-energy molecular orbitals first and by observing the Pauli principle. The ground state is obtained by following Hund's rule. The energy-level scheme for H_2^+ shown in Figure 3–5 is appropriate for diatomic molecules that have $1s$ valence orbitals.

The hydrogen molecule-ion, H_2^+, has the electronic structure $(\sigma^b)^1$; that is, it has one electron in a σ bonding orbital. Because there is one unpaired electron H_2^+ is paramagnetic. Notice that H_2^+ has one half an electron-pair bond, or one half a σ bond. The hydrogen molecule, H_2, has two electrons in the bonding orbital, thus it has the electronic structure $(\sigma^b)^2$. Therefore H_2 has one net bond and is diamagnetic. The helium molecule-ion, He_2^+, which has been detected during electric discharges through helium gas, has three electrons in the molecular-orbital system. Two electrons occupy the bonding orbital and one electron occupies the high-energy antibonding orbital; thus the structure is $(\sigma^b)^2(\sigma^*)^1$. From this structure we predict one half a bond for He_2^+. For He_2 the electronic configuration is $(\sigma^b)^2(\sigma^*)^2$ from which we predict no net bonding for diatomic helium. In general, if the $n = 1$ shells are filled for both atoms of a diatomic molecule, there is no net bonding from the four $1s$

electrons. Thus bonding (or the lack of bonding) would depend on the electrons in higher-energy orbitals.

A comparison of the molecular-orbital structures with bond energies and bond lengths is given in Table 3–1. The hydrogen molecule-ion, H_2^+, has a bond length of 1.06 Å, and H_2 has a bond length of 0.74 Å. The helium molecule-ion, He_2^+, exists and has a bond length of 1.08 Å. The helium molecule, He_2, does not exist, which is consistent with MO theory from which we predict no net bonds. The bond energies are 61 kcal mole^{-1} for the hydrogen molecule-ion, 103 kcal mole^{-1} for the hydrogen molecule, and 60 kcal mole^{-1} for the helium molecule-ion. We see from these examples that the molecular-orbital electronic configuration is useful for interpreting bond properties in molecules.

Table 3–1. Comparison of some molecular-orbital structures, net bonding electrons, bond lengths, and bond energies

Molecule	Bonding electrons	Antibonding electrons	Net bonding electrons	Bond length, Å	Experimental bond energy, kcal mole^{-1}
He_2	2	2	0	—[a]	—[a]
H_2^+	1	0	1	1.06	61
He_2^+	2	1	1	1.08	60
H_2	2	0	2	0.74	103

[a] Molecule is unstable.

3–2 MOLECULES WITH *s* AND *p* VALENCE ATOMIC ORBITALS

Now we will discuss diatomic molecules that have 2*s* and 2*p* valence atomic orbitals. Our purpose is to analyze carefully which orbitals are involved in electron-pair bonds in a molecule such as N_2. We will continue to follow the method of making linear (i.e., additive and subtractive) combinations of appropriate atomic orbitals. We know from atomic structure that 2*s* orbitals have less energy than 2*p* orbitals have. Thus in Figure 3–6 we see that there are three 2*p* orbitals of equal energy and one 2*s* orbital of less energy for each atom. We want to find the relative energies of the molecular orbitals for this system and to show how we build the electronic structures for diatomic molecules involving 2*s* and 2*p* valence orbitals.

Sigma orbitals

First consider the atomic 2*s* orbitals. The two molecular orbitals derived from 2*s* orbitals are similar to the H_2 molecular orbitals. The combination $2s_a + 2s_b$ concentrates electron density between the nuclei and is a σ bonding orbital. The combination $2s_a - 2s_b$ produces the relatively high-energy orbital that has

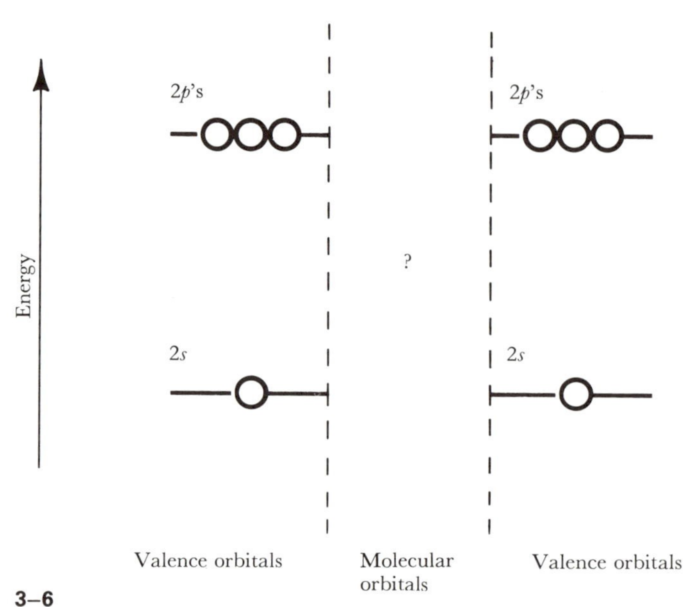

3–6
The relative energies of two sets of the 2*s* and 2*p* atomic orbitals of elements in the second row of the periodic table.

a nodal plane in which there is no probability of finding the electron. Electron density is forced out of the bonding region, thus this orbital is σ antibonding or σ^*. The σ^b and σ^* molecular orbitals are shown in Figure 3–7.

The three 2*p* orbitals are directed along the coordinate axes *X*, *Y*, and *Z*. The line that connects the nuclei in a diatomic molecule commonly is designated the *Z* axis. The two sets of corresponding *X* and *Y* axes are parallel and the *Z* axis is common to both nuclei. As shown in Figure 3–7, there are two different types of *p* orbitals in a diatomic molecule. One *p* orbital of each atom is aligned along the internuclear axis and is called the p_z orbital. In other words, the two $2p_z$ orbitals are directed toward each other and overlap along the *Z* axis. The other *p* orbitals are not directed toward each other in this way. The two p_x and the two p_y orbitals do not overlap along the *Z* axis, rather they overlap above and below it.

3–7▶
The six different kinds of molecular orbitals formed from the *s*, p_x, p_y, and p_z orbitals of two equivalent atoms in a diatomic molecule. The line drawn through the two nuclei is chosen as the *Z* axis. Plus and minus signs represent only the signs of the wave function, not electrical charge. The atomic orbitals from which these are obtained are shown, with their appropriate signs, at the upper right of each molecular orbital. The atomic orbitals used are *s* (top row), p_z (middle row), and p_x (bottom row), or the equivalent p_y. Bonding orbitals are in the left column; antibonding orbitals are in the right column. Dashed lines with the letter "*o*" represent nodal planes of zero electron density.

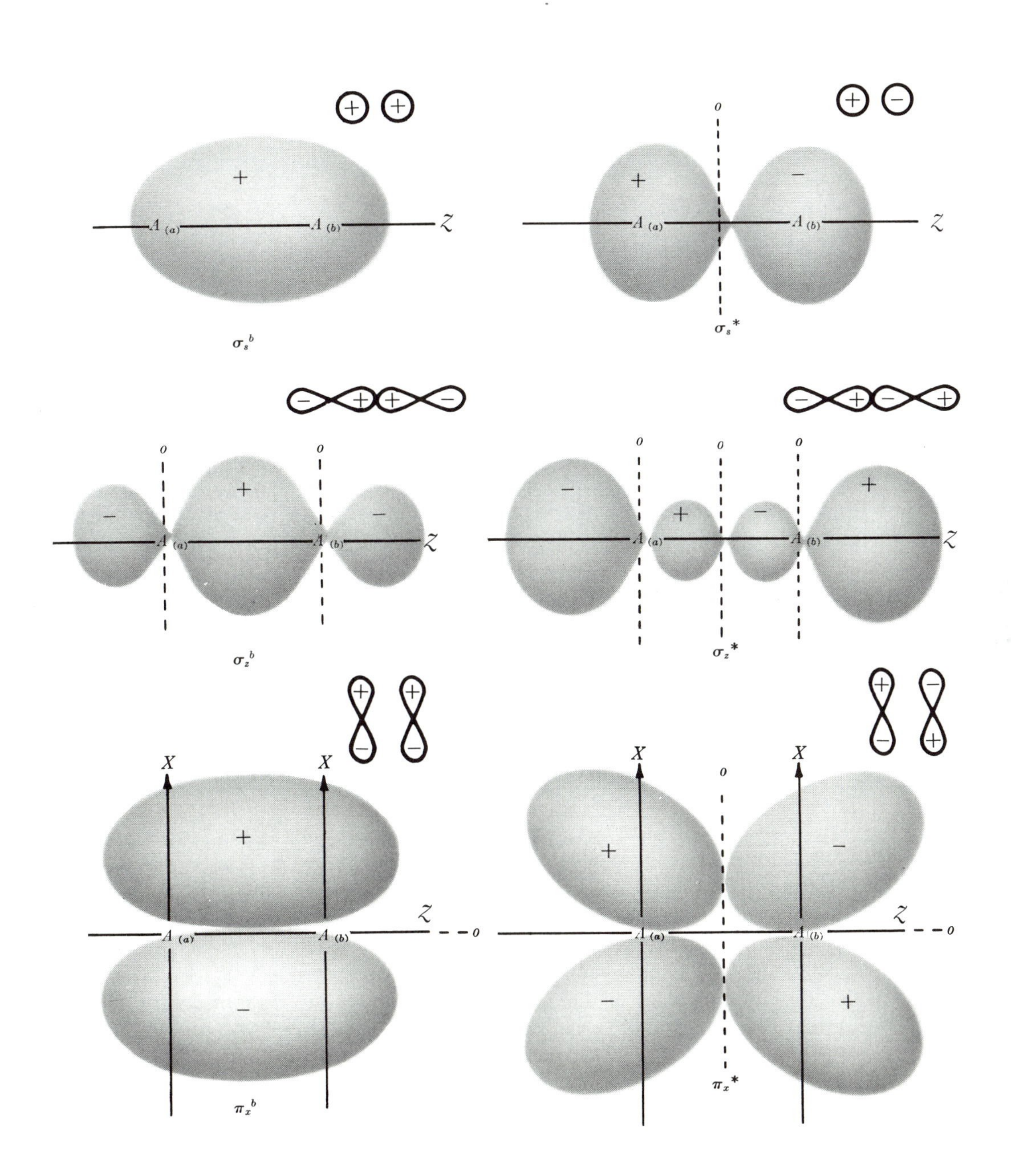
+
A (a)
A (b)
Z
$\sigma_s{}^b$
0
+
−
A (a)
A (b)
Z
$\sigma_s{}^*$
0
−
+
−
A (a)
A (b)
Z
0
$\sigma_z{}^b$
0
−
+
−
+
A (a)
A (b)
Z
$\sigma_z{}^*$
X
X
+
Z
A (a)
A (b)
0
−
$\pi_x{}^b$
X
X
0
+
−
Z
A (a)
A (b)
0
−
+
$\pi_x{}^*$

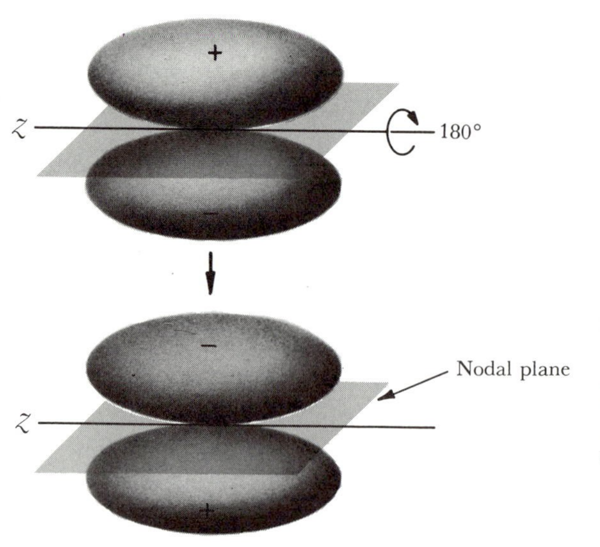

3–8
The overlap of two $2p_x$ orbitals gives rise to a π molecular orbital. A π molecular orbital changes sign when rotated 180° around the internuclear axis, but the electron distribution remains unchanged.

The combination of $z_a + z_b$ is of the same type as $s_a + s_b$; that is, the orbital is symmetrical around the Z axis.[1] Therefore it is a σ molecular orbital. Furthermore, electron density is greater in the overlap region, which means that $z_a + z_b$ is a bonding orbital. To distinguish $z_a + z_b$ from $s_a + s_b$ we add the subscript z to σ^b to indicate that the molecular orbital is made from p_z valence orbitals. (We add the subscript s to molecular orbitals constructed from s valence orbitals.) Thus σ_z^b is the bonding molecular orbital made from two $2p_z$ atomic orbitals. The other combination of $2p_z$ orbitals is the minus combination, $z_a - z_b$, which reduces electron density in the overlap region. As in $s_a - s_b$ there is a nodal plane halfway between the nuclei. Thus the $z_a - z_b$ orbital is antibonding and is denoted σ_z^*. The σ_z^b and σ_z^* molecular orbitals are shown in Figure 3–7.

Pi orbitals

Now we investigate the molecular orbitals formed from the $2p_x$ and $2p_y$ orbitals. The combination $x_a + x_b$ is a new type of molecular orbital. Since p orbitals have a node at the nucleus, the molecular orbital $x_a + x_b$ has a nodal plane that contains the internuclear line. Thus if we rotate the molecular orbital $x_a + x_b$ by 180° it changes sign, thereby becoming $-x_a - x_b$ (Figure 3–8). This type of molecular orbital is called a *pi* or *π orbital*. Pi molecular orbitals originate from combinations of parallel p orbitals that overlap above and below

[1] We will abbreviate $2p_{z_a}$ as z_a, $2p_{z_b}$ as z_b, $2p_{x_a}$ as x_a, and so on.

the internuclear line. The $x_a + x_b$ orbital concentrates electron density in the bonding region, thus it is a π bonding or π^b orbital. To complete the designation we say it is π_x^b, that is, a combination of the two $2p_x$ atomic orbitals. It should be clear that an equivalent $y_a + y_b$ orbital exists, which we call π_y^b. Thus there are two bonding π combinations that have the same shape and the same energy, and their orientations in space are mutually perpendicular.

Now let us consider the combinations $x_a - x_b$ and $y_a - y_b$, which have a nodal plane containing the internuclear axis. They are π antibonding orbitals (π^*) because they have reduced electron density (and an additional nodal plane perpendicular to the internuclear axis) between the nuclei. The π_x^b and π_x^* molecular orbitals are shown in Figure 3–7.

To summarize, we started with eight valence orbitals (one 2*s* and three 2*p* orbitals on each atom) and constructed eight molecular orbitals: σ_s^b, σ_s^*, σ_z^b, σ_z^*, π_x^b, π_y^b, π_x^*, and π_y^*. The relative energies of the molecular orbitals can be obtained from calculations and from experiments. Experimental data that are particularly helpful are obtained from absorption and emission spectroscopic measurements. In building the electronic structures of homonuclear (the same kind of nuclei) diatomic molecules, we know that the molecular orbitals are occupied in the following order of increasing energy:

$$\sigma_s^b < \sigma_s^* < \pi_x^b = \pi_y^b < \sigma_z^b < \pi_x^* = \pi_y^* < \sigma_z^*$$

This energy-level scheme is shown in Figure 3–9(a). An alternative energy-level ordering, which probably is correct for the homonuclear diatomic molecules O_2 and F_2, is shown in Figure 3–9(b). In each scheme there are six energy levels, which can accommodate a total of 16 electrons.

s–p Sigma hybridization

For a small energy separation between 2*s* and 2*p* orbitals, such as indicated in Figure 3–9(a), the σ_s^b molecular orbital in a diatomic molecule, A_2, almost certainly will mix, or *hybridize*, with the two $2p_z$ valence orbitals that are oriented along the bond axis. Such a hybridization of 2*s* and 2*p* valence orbitals in molecular orbital formation is favorable because, as shown in Figures 3–10 and 3–11, it increases the valence-orbital overlap between the nuclei. As a result, the electron density increases in this region and the energy of the σ_s^b molecular orbital is decreased. [Compare the MO diagram that includes *s–p* hybridization, Figure 3–9(a), with the diagram that neglects it, Figure 3–9(b).]

The value of λ in the hybrid wave function $2s + \lambda 2p_z$ (Figure 3–10) for any particular diatomic molecule is determined by the degree of hybridization that gives the largest net increase in the stability (decrease in the energy) of the σ_s^b bonding orbital. There are two important factors that determine the degree of *s–p* hybridization. As λ increases from zero to one the total overlap continually increases. But remember that the 2*p* orbital has more energy than

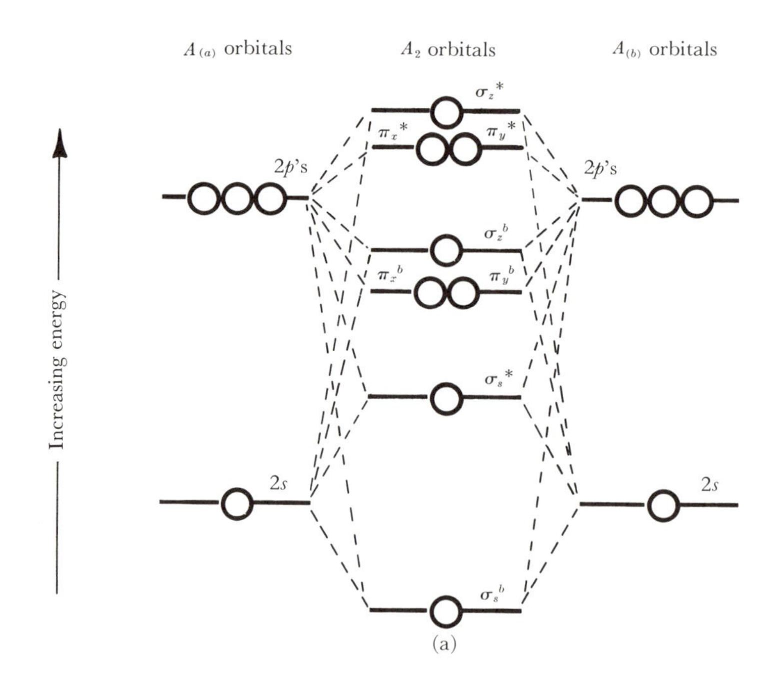

3–9
Molecular-orbital energy-level diagrams for homonuclear diatomic molecules. (a) Relative energies of molecular orbitals in the case of appreciable *s–p* hybridization. This level scheme has been established for most homonuclear diatomic molecules through detailed experiments involving magnetic and spectroscopic properties of molecules. The degree of *s–p* hybridization becomes smaller as the energy separation of the valence *s* and *p* atomic levels becomes larger. (b) Suggested relative order of molecular orbitals in systems with negligible *s–p*

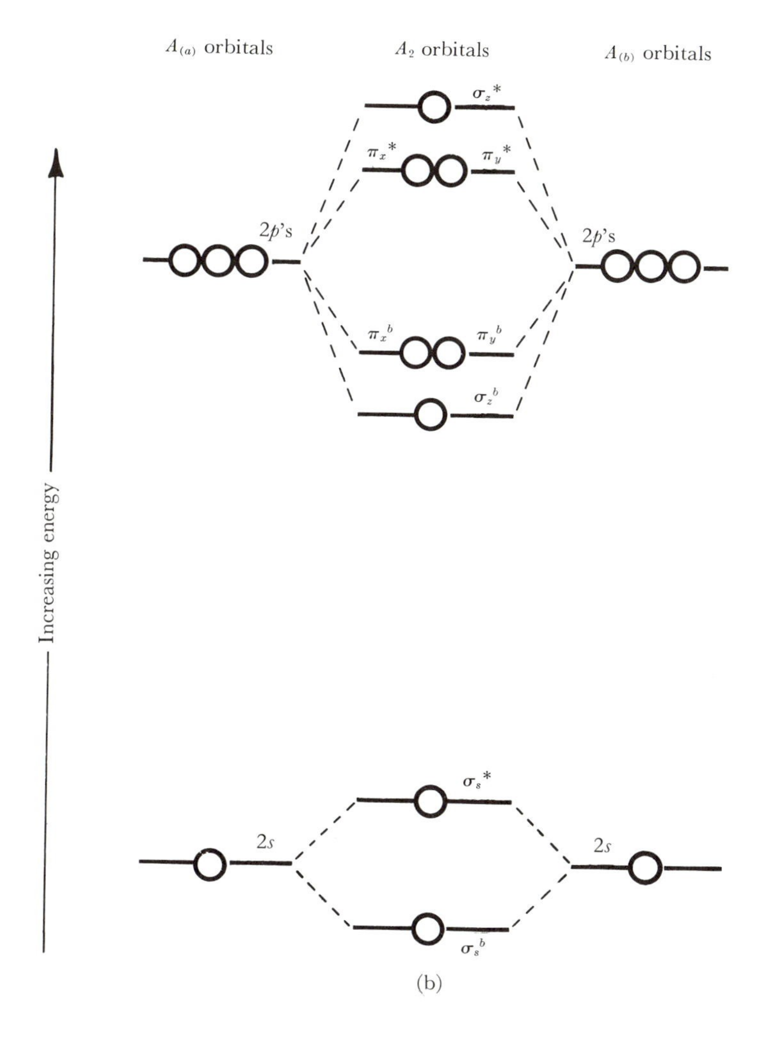

(b)

hybridization. This order of energy levels should be observed when the energy separation of the atomic *s* and *p* valence levels is relatively large. Referring to Table 3–3 we see that the large $2s$–$2p$ energy separation in F atoms makes the F_2 molecule a likely candidate for this scheme of energy levels. Additionally, there is some spectroscopic evidence that the electronic energy levels of the O_2 molecule are consistent with this diagram.

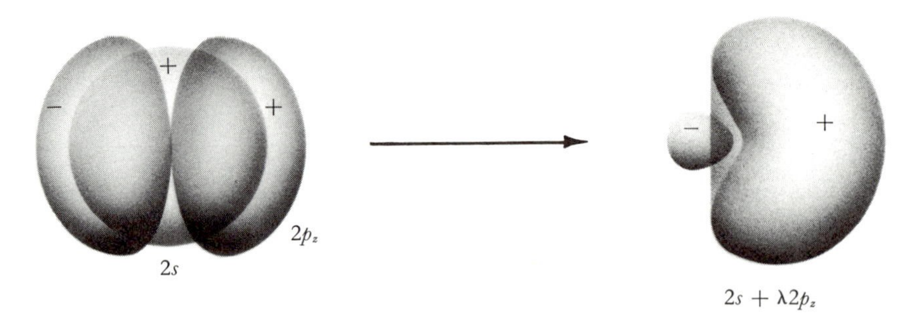

3–10
Hybridization of $2s$ and $2p_z$ valence orbitals. The $2s + 2p_z$ combination has a very large overlap, thus this combination makes a bonding molecular orbital of lower energy than the combination of two $2s$ orbitals.

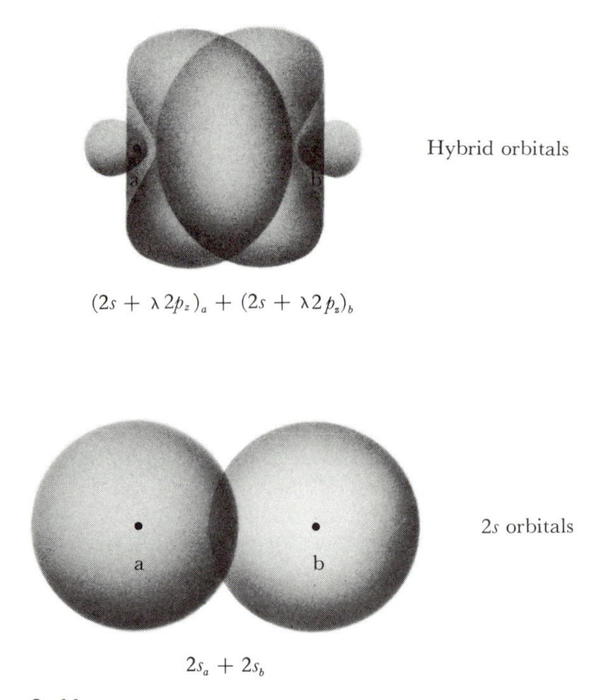

3–11
Comparison of overlap between two $2s$ orbitals and two *sp* hybrid orbitals. There is more overlap between the hybrid orbitals $(2s + \lambda 2p_z)_a$ and $(2s + \lambda 2p_z)_b$ than between the $2s$ orbitals. The overlap is greatly exaggerated for emphasis.

the 2s, so the energy of the isolated sp hybrid orbital also increases as λ increases from zero to one. Calculations can be performed to determine the optimum value of λ in a particular case. For example, in the Li_2 molecule the calculated λ value is approximately 0.2. Even with this relatively small degree of hybridization the calculated bond energy of Li_2 is significantly larger than that calculated for pure 2s bonding.

3–3 HOMONUCLEAR DIATOMIC MOLECULES

Now we are able to discuss the electronic structures and bonding properties of homonuclear diatomic molecules of elements in the second row of the periodic table. The bond lengths and bond energies of some important homonuclear diatomic molecules and ions are listed in Table 3–2.

Lithium

A lithium atom has one 2s valence electron. In Li the 2s–2p energy difference is small and the σ_s^b molecular orbital of Li_2 undoubtedly has considerable 2p character. The two valence electrons in Li_2 occupy the σ_s^b molecular orbital, thereby giving the ground-state electronic configuration $(\sigma_s^b)^2$. In agreement with theory, experimental measurements show that a lithium molecule has no unpaired electrons. With two electrons in a bonding molecular orbital there is one net bond. The bond length of Li_2 is 2.67 Å, compared with 0.74 Å for H_2. The longer bond in Li_2 is consistent with the larger effective radius of the Li 2s orbital. The mutual repulsion of the two 1s electron pairs, an interaction not present in H_2, also is partly responsible for the longer bond in Li_2.

The bond energies of H_2 and Li_2 are 103 kcal mole^{-1} and 26.3 kcal mole^{-1}, respectively. The two σ_s^b electrons on the average are much farther from the shielded nuclei in Li_2 than from the nuclei in H_2. At the longer distances the potential energy due to electron–nucleus attractions is smaller, thus the two σ_s^b $(2s_a + 2s_b)$ electrons do not bind Li_2 as strongly as the two σ_s^b $(1s_a + 1s_b)$ electrons do in H_2.

Beryllium

A beryllium atom has the valence electronic structure $2s^2$. The electronic configuration of Be_2 would be $(\sigma_s^b)^2(\sigma_s^*)^2$. This configuration gives no net bonds $[(2 - 2)/2 = 0]$, which is consistent with the absence of Be_2 from the family of stable second-row diatomic molecules.

Boron

Atomic boron has the valence electronic configuration $2s^22p^1$. For B_2 there are six valence electrons to assign to molecular orbitals. With the ordering of levels shown in Figure 3–9(a) the ground-state electronic configuration of B_2

Table 3–2. Bond properties of some homonuclear diatomic molecules and ions[a]

Molecule	Bond length, Å	Bond dissociation energy, kcal $mole^{-1}$
Ag_2	—	38.7 ± 2.2
As_2	2.288	91.3
Au_2	2.472	53.9 ± 2.2
B_2	1.589	65.5 ± 5
Bi_2	—	46.6 ± 1.5
Br_2	2.2809	45.440 ± 0.003
C_2	1.2425	144
Cl_2	1.988	57.18 ± 0.006
Cl_2^+	1.8917	99.2
Cs_2	—	10.4
Cu_2	2.2195	47.3 ± 2.2
F_2	1.417	33.2 ± 1.6
Ge_2	—	65
H_2	0.74116	103.24
H_2^+	1.06	61.06
He_2^+	1.080	77.0
I_2	2.6666	35.55
K_2	3.923	11.8
Li_2	2.672	26.3
N_2	1.0976	225.07
N_2^+	1.116	201.28
Na_2	3.078	17.3
O_2	1.20741	117.96
O_2^+	1.1227	—
O_2^-	1.26	93.9
O_2^{2-}	1.49	—
P_2	1.8937	114
Pb_2	—	23
Rb_2	—	11.3
S_2	1.889	100.69
Sb_2	2.21	71.3
Se_2	2.1663	77.6
Si_2	2.246	75
Sn_2	—	46
Te_2	2.5574	62.3

[a] The values of bond dissociation energy (*DE*) given generally refer to the ΔE° for the process $A_2(g) \rightarrow A(g) + A(g)$.

is $(\sigma_s^b)^2(\sigma_s^*)^2(\pi_x^b)^1(\pi_y^b)^1$, thereby giving one net bond. Both Li_2 and B_2 have one net bond, but the B–B bond length is 1.59 Å, which is much shorter than the Li–Li length of 2.67 Å. The shorter bond length in B_2 is a consequence of the decrease in effective atomic radius that accompanies the increase in Z_{eff} from Li to B (Section 2–2). The proposed electronic configuration of B_2 is supported by experimental measurements, which show that B_2 has two unpaired electrons in π-type orbitals.

The bond energy of B_2 is 65 kcal mole^{-1}, which is considerably larger than 26 kcal mole^{-1} for Li_2. These data also are consistent with the smaller atomic size of boron and the fact that its valence electrons are bound more firmly than the 2s electron of Li.

Carbon

Atomic carbon has the valence electronic configuration $2s^2 2p^2$. For C_2 there are eight valence electrons and the ground-state structure is $(\sigma_s^b)^2(\sigma_s^*)^2(\pi_{x,y}^b)^4$, which has no unpaired electrons. However, the σ_z^b orbital energy is greater than that of $\pi_{x,y}^b$ by about the energy required to pair two electrons. Thus the structure $(\sigma_s^b)^2(\sigma_s^*)^2(\pi_{x,y}^b)^3(\sigma_z^b)^1$, with two unpaired electrons, does not differ significantly in energy from the $(\sigma_s^b)^2(\sigma_s^*)^2(\pi_{x,y}^b)^4$ structure. The two structures can be pictured as

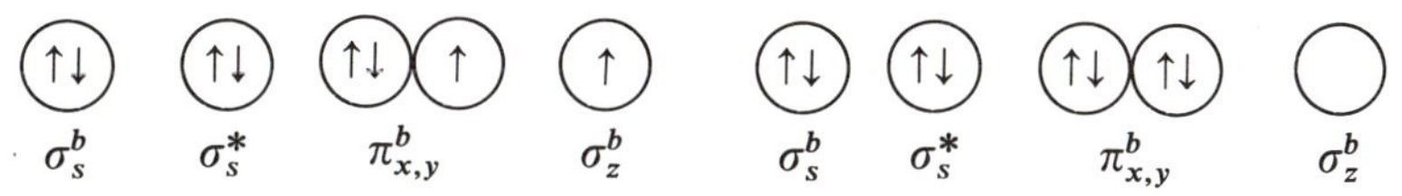

The most recent experiments indicate that the $(\sigma_s^b)^2(\sigma_s^*)^2(\pi_{x,y}^b)^4$ structure is lower in energy than the electronic state represented by the other configuration. The C_2 molecule absorbs light in the visible region of the spectrum, with maximum absorption at approximately 19,300 cm^{-1}. This absorption has been assigned to an electronic transition from the $\pi_{x,y}^b$ orbital to the σ_z^b orbital, abbreviated $\pi_{x,y}^b \rightarrow \sigma_z^b$. There are two net bonds predicted for C_2, which is compatible with the experimentally determined bond energy of 144 kcal mole^{-1} and the bond length of 1.24 Å.

Nitrogen

Atomic nitrogen has the valence electronic configuration $2s^2 2p^3$. The ground-state electronic configuration of N_2 is $(\sigma_s^b)^2(\sigma_s^*)^2(\pi_{x,y}^b)^4(\sigma_z^b)^2$, which is consistent with the observed diamagnetism of this molecule. The nitrogen molecule has three net bonds (one σ and two π), which is the maximum for a second-row diatomic molecule. The triple bond in N_2 accounts for its unusual stability, its extraordinarily large bond energy of 225 kcal mole^{-1}, and its very short bond length of 1.10 Å. Recall that the Lewis structure of N_2 is :N≡N:, with three bonds and two lone pairs. The Lewis structure now can be analyzed in terms of molecular orbital theory. The two lone pairs correspond to the $(\sigma_s^b)^2$ and $(\sigma_s^*)^2$ configurations; that is, an equal number of bonding and anti-bonding electron pairs results in no net bonds. The three bonds in the Lewis structure correspond to the two π bonds, $(\pi_{x,y}^b)^4$, and one σ bond, $(\sigma_z^b)^2$, proposed from MO theory.

The low-energy electronic transition $\pi^b_{x,y} \rightarrow \sigma^b_z$, which is observed for the C_2 molecule, is not possible in N_2 because σ^b_z is occupied fully in the ground state. As a result N_2 does not absorb light in the visible region. The lowest electronic transition in N_2, $\sigma^b_z \rightarrow \pi^*_{x,y}$, occurs at very high energy (approximately 70,000 cm^{-1}).

Oxygen

Atomic oxygen has the valence electronic configuration $2s^2 2p^4$. The electronic configuration of O_2 is $(\sigma^b_s)^2(\sigma^*_s)^2(\sigma^b_z)^2(\pi^b_{x,y})^4(\pi^*_x)^1(\pi^*_y)^1$. Following Hund's rule the electrons in $\pi^*_{x,y}$ have the same spin in the ground state, thereby giving two unpaired electrons in O_2. The oxygen molecule is paramagnetic, which is in agreement with this electronic structural model. In this respect the molecular orbital theory is superior to the simple Lewis picture $:\ddot{O}{=}\ddot{O}:$, which does not show that O_2 has two unpaired electrons.

Two net bonds (one σ and one π) are predicted for O_2. The bond energy of O_2 is 118 kcal $mole^{-1}$ and the bond length is 1.21 Å. The change in bond length caused by changing the number of electrons in the $\pi^*_{x,y}$ level of the O_2 system is instructive. The precise bond length of O_2 is 1.2074 Å. When an electron is added to the $\pi^*_{x,y}$ level of O_2, thereby forming O_2^-, the bond length increases to 1.26 Å. Addition of a second electron to form O_2^{2-} increases the bond length to 1.49 Å. This is in agreement with the prediction of 1 1/2 bonds for O_2^- and 1 bond for O_2^{2-}.

Fluorine

Atomic fluorine has the valence electronic configuration $2s^2 2p^5$. The electronic configuration of F_2 is $(\sigma^b_s)^2(\sigma^*_s)^2(\sigma^b_z)^2(\pi^b_{x,y})^4(\pi^*_{x,y})^4$, with no unpaired electrons and one net bond. This electronic structure is consistent with the diamagnetism of F_2, its 33 kcal $mole^{-1}$ bond energy, and the F–F bond length of 1.42 Å.

Neon

A neon atom has the closed-shell electronic configuration $2s^2 2p^6$. The hypothetical Ne_2 molecule would have the configuration $(\sigma^b_s)^2(\sigma^*_s)^2(\sigma^b_z)^2(\pi^b_{x,y})^4(\pi^*_{x,y})^4(\sigma^*_z)^2$ and zero net bonds. To date there is no experimental evidence for the existence of a stable diatomic neon molecule.

In summary, the MO theory provides an excellent framework for the correlation of bond lengths and bond energies of diatomic molecules. We already have shown for the series H_2^+, H_2, He_2^+, and He_2, that as the bond order increases, the bond length decreases and the bond energy increases. This relationship holds for comparisons of the bonds between atoms of approximately the same effective size. The quantitative correlations for B_2, C_2, N_2, O_2, and F_2 are:

Bond orders (number)

$$B_2\,(1) < C_2\,(2) < N_2\,(3) > O_2\,(2) > F_2\,(1)$$

Bond energies (kcal mole^{-1})

$$B_2\,(65) < C_2\,(144) < N_2\,(225) > O_2\,(118) > F_2\,(33)$$

Bond lengths (Å)

$$B_2\,(1.59) > C_2\,(1.24) > N_2\,(1.10) < O_2\,(1.21) < F_2\,(1.42)$$

3–4 HOMONUCLEAR DIATOMIC MOLECULES WITH $n > 2$ VALENCE ORBITALS

With proper adjustment of the principal quantum number of the valence orbitals, the molecular-orbital energy-level diagrams shown in Figure 3–9 for second-row diatomic molecules can be used to describe the electronic structures of homonuclear diatomic molecules of elements of the third row and higher.

Na_2, K_2, Rb_2, and Cs_2

The alkali-metal diatomic molecules all have the ground-state configuration $(\sigma_s^b)^2$ and one bond, and all are diamagnetic. The bond lengths and bond energies of Li_2, Na_2, K_2, Rb_2, and Cs_2 are given in Table 3–2. The bond lengths increase and the bond energies decrease regularly from Li_2 to Cs_2. The increase in bond lengths (consequently the decrease in bond energies) corresponds to the increase in the effective size of the atoms from Li to Cs.

Cl_2, Br_2, AND I_2

The ground-state electronic configuration of the halogen molecules is $(\sigma_s^b)^2(\sigma_s^*)^2(\pi_{x,y}^b)^4(\sigma_z^b)^2(\pi_{x,y}^*)^4$, thereby indicating one net bond. The molecules are diamagnetic. Table 3–2 gives bond lengths and bond energies for F_2, Cl_2, Br_2, and I_2. The bond lengths increase predictably from F_2 to I_2, but the bond energies are irregular, increasing from F_2 to Cl_2 then decreasing from Cl_2 to I_2. We can explain the irregularity in bond energies for the halogens as follows. The molecular orbital structure $(\pi_{x,y}^b)^4(\pi_{x,y}^*)^4$ is equivalent to four $2p(\pi)$ lone pairs:

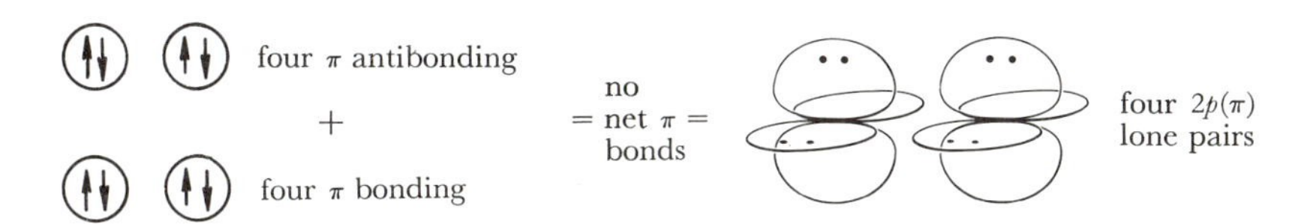

Repulsion of the electrons in $2p(\pi)$ orbitals on adjacent halogen atoms is a factor that is not present in the alkali-metal diatomic molecules. Due to the very small size of F, this interelectronic repulsion is unusually large in F_2, thereby giving rise to an anomalously small bond energy.

3–5 HYDROGEN FLUORIDE MOLECULE

We use hydrogen fluoride, HF, as an example of a heteronuclear diatomic molecule for a detailed molecular orbital treatment. When combining valence orbitals of different atoms it is helpful to know the relative energies of the orbitals. Table 3–3 gives valence-orbital ionization energies for the atoms hydrogen through krypton. The corresponding energies of the valence-electron orbitals, which we use to construct MO diagrams, are obtained simply by taking the negative of the *IE*'s.

A hydrogen atom has a 1*s* valence orbital and a fluorine atom has 2*s* and 2*p* valence orbitals. The orbital energies are

$$\text{hydrogen: } 1s,\ -110{,}000\ \text{cm}^{-1}$$
$$\text{fluorine: } 2s,\ -374{,}000\ \text{cm}^{-1};\ 2p,\ -151{,}000\ \text{cm}^{-1}$$

Table 3–3. Valence-orbital ionization energies[a] in units of 10^3 cm^{-1}

Atom	1*s*	2*s*	2*p*	3*s*	3*p*	4*s*	4*p*
H	110	—	—	—	—	—	—
He	198	—	—	—	—	—	—
Li	—	44	—	—	—	—	—
Be	—	75	—	—	—	—	—
B	—	113	67	—	—	—	—
C	—	157	86	—	—	—	—
N	—	206	106	—	—	—	—
O	—	261	128	—	—	—	—
F	—	374	151	—	—	—	—
Ne	—	391	174	—	—	—	—
Na	—	—	—	42	—	—	—
Mg	—	—	—	62	—	—	—
Al	—	—	—	91	48	—	—
Si	—	—	—	121	63	—	—
P	—	—	—	151	82	—	—
S	—	—	—	167	94	—	—
Cl	—	—	—	204	111	—	—
Ar	—	—	—	236	128	—	—
K	—	—	—	—	—	35	—
Ca	—	—	—	—	—	49	—
Zn	—	—	—	—	—	76	—
Ga	—	—	—	—	—	102	48
Ge	—	—	—	—	—	126	61
As	—	—	—	—	—	142	73
Se	—	—	—	—	—	168	87
Br	—	—	—	—	—	194	101
Kr	—	—	—	—	—	222	115

[a] The reference zero point is the ionized atom. Thus the corresponding valence-orbital energies are obtained simply by changing the sign of the *IE*. For example, the 1*s* orbital energy of atomic H is $-110{,}000$ cm^{-1}.

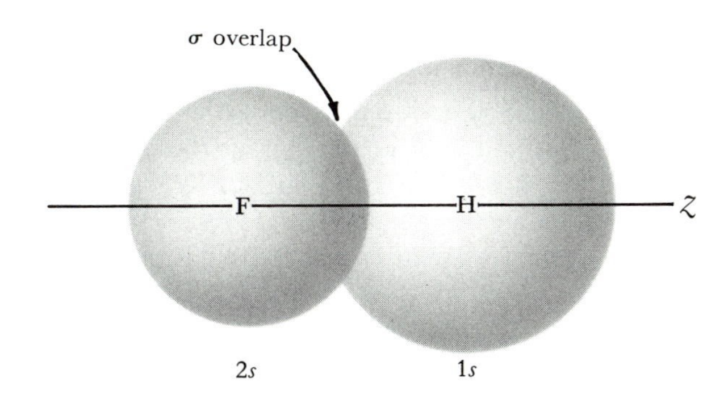

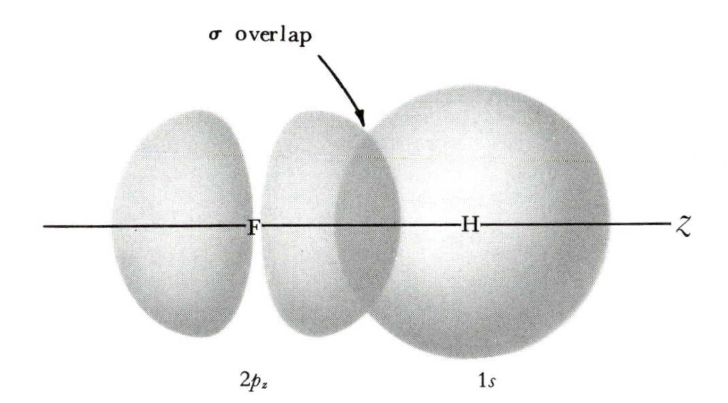

3–12
Overlap of the hydrogen 1s orbital with the valence orbitals of fluorine. The net overlap of a 2p_x or 2p_y orbital of fluorine with the hydrogen 1s orbital is zero, and these two p orbitals cannot be used in forming molecular orbitals.

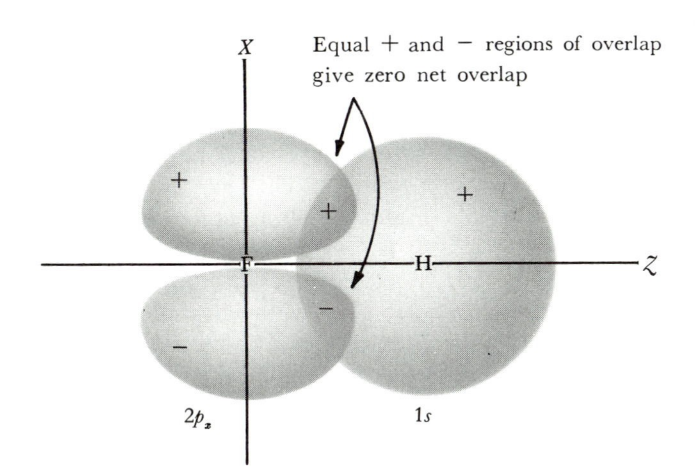

Since the $2p$ orbitals in fluorine are much closer in energy to the $1s$ orbital in hydrogen we will consider, as a first approximation, only the interaction of the $1s$ orbital of hydrogen with the $2p$ orbitals of fluorine in molecular-orbital formation. The overlap of the hydrogen $1s$ orbital with the fluorine $2p$ orbitals is shown in Figure 3–12. The Z axis corresponds to the internuclear line, and we see that the $1s$(H) and $2p_z$(F) orbitals overlap. Therefore the bonding orbital is represented by the combination of these orbitals:

$$\sigma^b = \lambda_1[1s(\mathrm{H})] + \lambda_2[2p_z(\mathrm{F})]$$

The coefficients λ_1 and λ_2 give the relative weights of the hydrogen $1s$ and the fluorine $2p_z$ orbitals in the σ^b molecular orbital. In this case the combining valence orbitals have different weights in the molecular orbital because they have different energies. As we discussed in Chapter 2, electrons naturally will be "pulled" toward the atom that has the larger electronegativity, which therefore furnishes the lower-energy valence orbital. In the molecular-orbital formulation for HF λ_2 is much greater than λ_1 in the bonding orbital, thereby giving significantly more weight to the lower-energy fluorine $2p_z$ orbital. Thus there are virtually two electrons in the fluorine $2p_z$ orbital in the ground state.

The antibonding orbital in HF can be written

$$\sigma^* = \gamma_1[1s(\mathrm{H})] - \gamma_2[2p_z(\mathrm{F})]$$

The antibonding molecular orbital is not occupied by electrons in HF in the ground state, but it could be occupied in certain excited electronic states. The coefficient γ_2 gives the relative weight of the fluorine $2p_z$ orbital in the antibonding molecular orbital. Electron density in σ^* will be reduced between the nuclei. Furthermore, since the bonding orbital has "used" most of the fluorine $2p_z$ orbital, electron density will be forced mainly into the domain of the hydrogen nucleus. This means that γ_2 will be considerably less than γ_1; that is, an electron would be associated with the higher-energy $1s$(H) orbital if it were found in the σ^* molecular orbital.

The fluorine $2p_x$ and $2p_y$ orbitals are suitable for π molecular orbitals. However, atomic hydrogen has only a $1s$ valence orbital, which is involved solely in σ bonding. The $1s$(H) orbital has zero *net* overlap with the $2p_x$ and $2p_y$ orbitals (Figure 3–12), thus the fluorine $2p_x$ and $2p_y$ orbitals are nonbonding in HF.

The molecular orbitals for HF and their relative energies are shown in Figure 3–13. The valence orbitals of fluorine are on the right in the diagram, with the $2p$ energy level above the $2s$ level. On the left, the hydrogen $1s$ energy level is placed higher than the fluorine $2p$ level, in agreement with their known relative energies. The σ^b and σ^* molecular orbitals are in the center. The σ^b molecular orbital is lower in energy than the fluorine $2p_z$ orbital, and the diagram illustrates that σ^b has a large component of the fluorine $2p_z$ orbital and a small

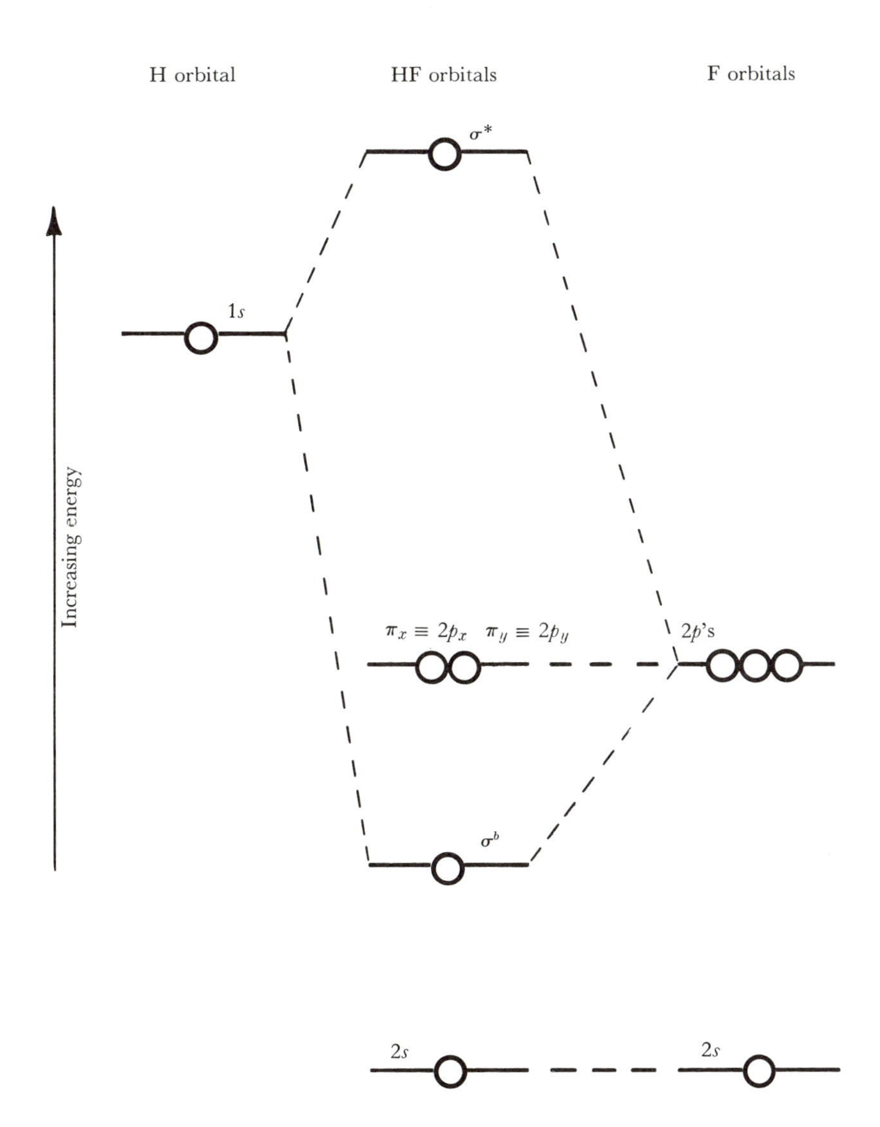

3–13
Relative energies of atomic and molecular orbitals in HF. The energy of an electron in atomic hydrogen 1*s* orbital is $-110{,}000\ \mathrm{cm}^{-1}$ (the first ionization energy of H is $+110{,}000\ \mathrm{cm}^{-1}$), and the energy in the 2*p* orbitals of F is $-151{,}000\ \mathrm{cm}^{-1}$ (first ionization energy of F is $+151{,}000\ \mathrm{cm}^{-1}$). Not drawn to scale because the 2*s* orbital actually is much lower in energy than shown.

component of the hydrogen $1s$ orbital. The σ^* molecular orbital is higher in energy than the hydrogen $1s$ orbital, and the diagram shows that σ^* is composed mainly of the hydrogen $1s$ orbital. The fluorine $2p_x$ and $2p_y$ orbitals are shown in the molecular orbital column as π-type molecular orbitals. They are nonbonding because hydrogen has no p valence orbitals.

There are eight valence electrons to place in the molecular orbitals for HF (Figure 3–13). Seven are from fluorine ($2s^22p^5$) and one is from hydrogen ($1s$). Therefore the ground-state electronic structure of HF is $(2s)^2(\sigma^b)^2(2p_x)^2(2p_y)^2$. In addition to the one σ bond there are three lone pairs, which corresponds to the Lewis structure (H—$\ddot{\underset{..}{F}}$:).

Since the electrons in the σ^b molecular orbital spend more time in the vicinity of the fluorine nucleus than of the hydrogen nucleus, it follows that there is a separation of charge in the ground state of HF. Specifically, hydrogen has a partial positive charge ($\delta+$) and fluorine has a partial negative charge ($\delta-$):

$$H^{\delta+}F^{\delta-}$$

An extreme situation would exist if both σ^b electrons spent *all* their time around the fluorine atom. In that case an HF molecule would be composed of H^+ and F^- ions. Recall that a molecule that can be formulated accurately as an "ion pair" is described as an ionic molecule. This situation is encountered in a diatomic molecule only if the valence orbital of one atom has much less energy than has the valence orbital of the other atom. An example is NaF. The difference between the ionization energies of the Na $3s$ orbital and the F $2p$ orbital is 109,000 cm^{-1}. In HF the energy difference between the H $1s$ orbital and the F $2p$ orbital is 41,000 cm^{-1}. Therefore the HF molecule is not as ionic as NaF, but we say that HF has partial ionic character.

Dipole moment of HF

A heteronuclear diatomic molecule such as HF has an electric dipole moment caused by a charge separation. This dipole moment is equal to the product of the charge and the distance of separation:

$$\text{dipole moment} = \mu = (qe)R$$

If the distance R is in centimeters and the charge qe is in electrostatic units, then μ is in esu cm. Since the unit of electronic charge (e) is 4.8×10^{-10} esu and bond distances are of the order of 10^{-8} cm (1 Å), dipole moments are of the order of 10^{-18} esu cm. It is convenient to express μ in debye units (D) (10^{-18} esu cm $=$ 1 debye). As a first approximation, if we consider the charges to be centered at each nucleus, then R is simply the equilibrium internuclear separation in the molecule.

Because it is possible to measure dipole moments we have an experimental method of estimating the partial ionic character of heteronuclear diatomic molecules. The observed dipole moment of HF is 1.82 D. For $R = 0.92$ Å (or 0.92×10^{-8} cm) the ionic structure H^+F^- has a calculated dipole moment of 4.4 D (4.8×10^{-10} esu $\times$ 0.92×10^{-8} cm). Thus the separated partial charge calculated from the dipole moment data is 1.82/4.4 = 0.41, which represents a partial ionic character of 41%.

Dipole moments for several diatomic molecules are given in Table 3–4.

Table 3–4. Dipole moments of some diatomic molecules

Molecule	Dipole moment, D
LiH	5.88
HF	1.82
HCl	1.08
HBr	0.82
HI	0.44
O_2	0
CO	0.112
NO	0.153
ICl	0.65
BrCl	0.57
ClF	0.88
BrF	1.29
KBr	10.41
KCl	10.27
KF	8.60
KI	11.05

3–6 A GENERAL *AB* HETERONUCLEAR DIATOMIC MOLECULE

Now we will describe the bonding in a general diatomic molecule *AB*, in which *B* has a larger electronegativity than *A*, and both *A* and *B* have *s* and *p* valence orbitals. The molecular-orbital energy levels for *AB* are illustrated in Figure 3–14. The *s* and *p* orbitals of *B* are placed lower than the *s* and *p* orbitals of *A*, in agreement with the electronegativity difference between *A* and *B*. The σ and π bonding and antibonding orbitals for *AB* are formed in the same manner as for A_2, but with the coefficients of the valence orbitals larger for *B* in the bonding orbitals and larger for *A* in the antibonding orbitals. This means that the electrons in the bonding orbitals spend more time near the more electronegative atom, *B*. In higher-energy antibonding orbitals the electrons spend more time near the less electronegative atom, *A*. Spatial representations of the molecular orbitals for a general *AB* molecule are given in Figure 3–15.

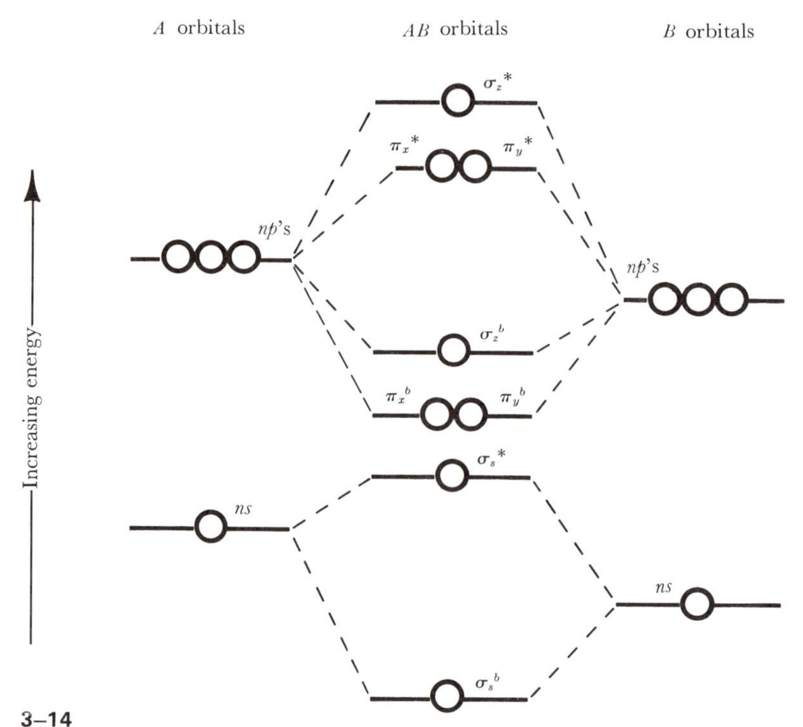

3–14
Relative orbital energies in a general *AB* molecule in which *B* is more electronegative than *A*.

The bond properties of several examples of the heteronuclear diatomic molecules and ions listed in Table 3–5 are discussed in the following paragraphs.

BN (eight valence electrons)

The ground-state electronic configuration for BN is $(\sigma_s^b)^2(\sigma_s^*)^2(\pi_{x,y}^b)^3(\sigma_z^b)^1$, from which we predict two bonds in the molecule. Thus the BN molecule is electronically similar to C_2, except that for BN the configuration with two unpaired electrons is more stable. The bond lengths of C_2 and BN are 1.243 Å and 1.281 Å, respectively. The BN bond energy of 92 kcal mole^{-1} is suspiciously low compared with 144 kcal mole^{-1} for C_2. Further experimental work is necessary to verify the BN bond energy.

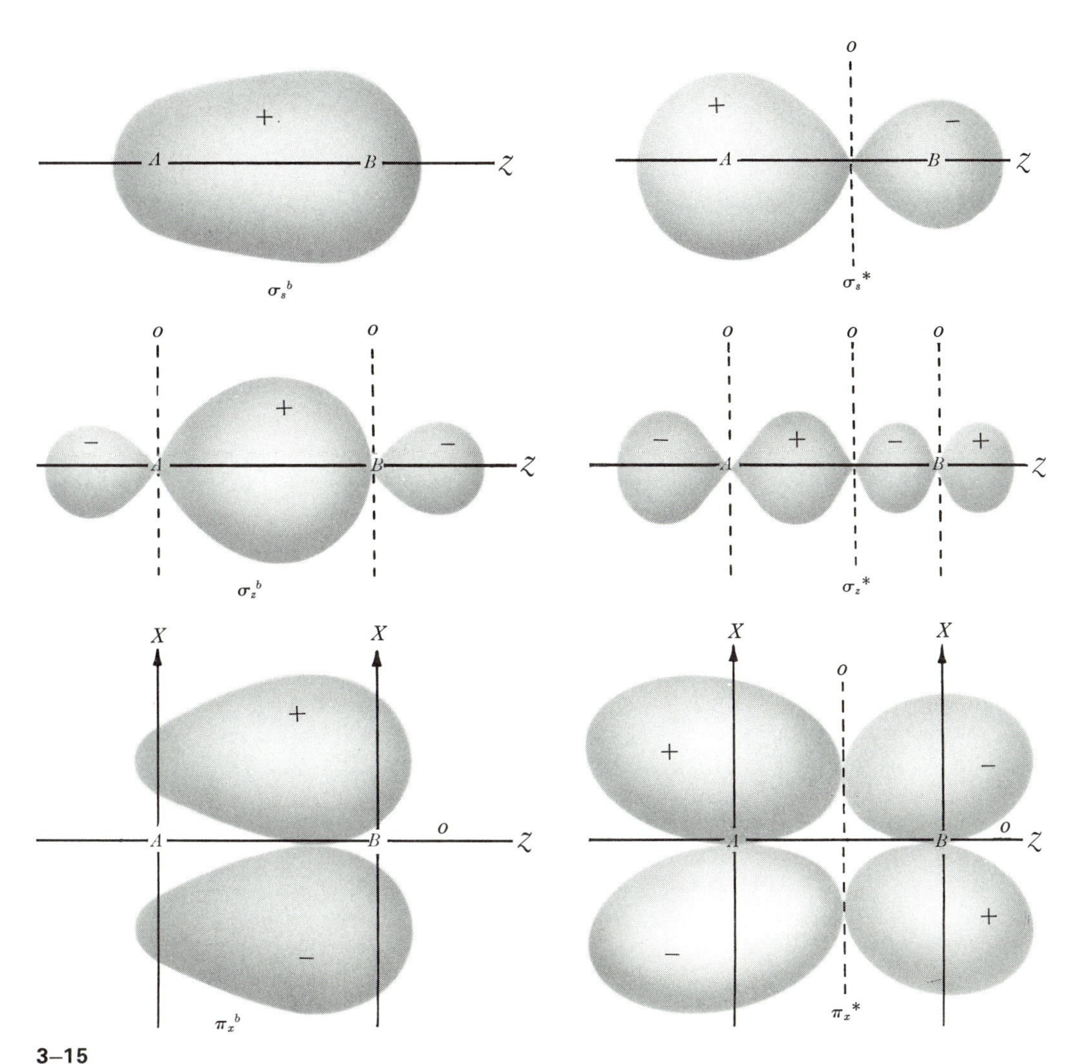

3–15
Spatial representation of molecular orbitals for an *AB* molecule in which *B* is more electronegative than *A*. Lines with the letter "*o*" represent nodal planes of zero electron density.

Table 3–5. Bond properties of some heteronuclear diatomic molecules and ions

Molecule	Bond length, Å	Bond dissociation energy, kcal mole^{-1}
AsN	1.620	115
AsO	1.623	113
BF	1.262	131
BH	1.2325	70
BN	1.281	92
BO	1.2043	191.2 ± 2.3
BaO	1.940	130.4 ± 6
BeF	1.3614	135.9 ± 2.3
BeH	1.297	53
BeO	1.3308	106.1 ± 2.3
BrCl	2.138	52.1
BrF	1.7555	55
CF	1.2718	106
CH	1.1202	80
CN	1.1719	188
CN^+	1.1727	—
CN^-	1.14	—
CO	1.1283	255.8
CO^+	1.1152	192.4
CP	1.5583	122.1 ± 5
CS	1.5349	173.6 ± 3.5
CSe	1.66	138 ± 5
CaO	1.822	91.32 ± 1.4
ClF	1.6281	60.3
CsBr	3.072	91.5
CsCl	2.9062	101.7
CsF	2.345	122
CsH	2.494	42
CsI	3.315	75.4
GeO	1.650	157
HBr	1.4145	86.5
HBr^+	1.459	—
HCl	1.2744	102.2
HCl^+	1.3153	108.3
HF	0.91680	135.1
HI	1.6090	70.5
HS	1.3503	81.4
IBr	2.485	41.90
ICl	2.32070	49.63
IF	1.908	45.7
KBr	2.8207	91.4
KCl	2.6666	100.8
KF	2.1715	118.9

Table 3–5 (*Continued*)

Molecule	Bond length, Å	Bond dissociation energy, kcal mole^{-1}
KH	2.244	43
KI	3.0478	77.2
LiBr	2.1704	101
LiCl	2.018	113.25
LiF	1.5639	135.8
LiH	1.5953	56
LiI	2.3919	81
MgO	1.749	81
NH	1.045	85
NH^+	1.081	—
NO	1.1508	162
NO^+	1.0619	—
NP	1.4910	—
NS	1.495	115
NS^+	1.25	—
NaBr	2.502	88
NaCl	2.3606	98.5
NaF	1.9260	113.9
NaH	1.8873	47
NaI	2.7115	69
NaK	—	14.3
NaRb	—	13.8
OH	0.9706	101.5
OH^+	1.0289	101.0
PH	1.4328	—
PN	1.4869	174.6
PO	1.473	124
RbBr	2.9448	90.9
RbCl	2.7868	102.8
RbF	2.2704	119.5
RbH	2.367	39
RbI	3.1769	77.7
SO	1.4810	123.66
SbO	1.848	74
SiF	1.6008	129.5
SiH	1.5201	74
SiN	1.575	104
SiO	1.5097	182.8
SiS	1.929	148
SnH	1.785	74
SnO	1.838	126.5
SnS	2.209	110.3
SrO	1.9199	99.2

BO, CN, and CO^+ (nine valence electrons)

The BO, CN, and CO^+ molecular species all have the ground-state configuration $(\sigma_s^b)^2(\sigma_s^*)^2(\pi_{x,y}^b)^4(\sigma_z^b)^1$. For each molecule 2 1/2 bonds are predicted. The bond lengths all are shorter than that of BN (or C_2), being 1.204 Å for BO, 1.172 Å for CN, and 1.115 Å for CO^+. The bond energies of the neutral molecules are greater than the bond energy of BN, being 191 kcal mole^{-1} for BO and 188 kcal mole^{-1} for CN.

NO^+, CO, and CN^- (ten valence electrons)

The NO^+, CO, and CN^- molecular species are isoelectronic with N_2. From the configuration $(\sigma_s^b)^2(\sigma_s^*)^2(\pi_{x,y}^b)^4(\sigma_z^b)^2$ we predict one σ and two π bonds. The bond lengths of NO^+, CO, and CN^- increase with increasing negative charge, being 1.062 Å for NO^+, 1.128 Å for CO, and 1.14 Å for CN^-. Comparing CO with BO and NO^+ with CO^+ (species with like charge), we see that the bond length of CO is less than that of BO and that of NO^+ is less than that of CO^+, as we should expect. The bond energy of CO is 255.8 kcal mole^{-1}, which is greater than the bond energy of 225 kcal mole^{-1} for N_2.

NO (eleven valence electrons)

The electronic configuration of NO is $(\sigma_s^b)^2(\sigma_s^*)^2(\pi_{x,y}^b)^4(\sigma_z^b)^2(\pi_{x,y}^*)^1$. Since the eleventh electron is in a π^* orbital the bond order is 2 1/2, which is one half less than for NO^+. The bond length of NO is 1.151 Å, which is longer than both the CO and NO^+ bonds. The bond energy of NO is 162 kcal mole^{-1}, which is considerably less than the bond energy of CO.

SUGGESTIONS FOR FURTHER READING

C. J. Ballhausen and H. B. Gray, *Molecular Orbital Theory*, Benjamin, Menlo Park, Calif., 1964.

E. Cartmell and G. W. A. Fowles, *Valency and Molecular Structure*, 3rd ed., Butterworths, London, 1966.

A. Companion, *Chemical Bonding*, McGraw-Hill, New York, 1964.

C. A. Coulson, *Valence*, 2nd ed., Oxford, New York, 1961.

H. B. Gray, *Electrons and Chemical Bonding*, Benjamin, Menlo Park, Calif., 1965.

G. Herzberg, *Spectra of Diatomic Molecules*, Van Nostrand, Princeton, N.J., 1950.

M. Karplus and R. N. Porter, *Atoms and Molecules: An Introduction for Students of Physical Chemistry*, Benjamin, Menlo Park, Calif., 1970.

G. C. Pimentel and R. D. Spratley, *Chemical Bonding Clarified Through Quantum Mechanics*, Holden-Day, San Francisco, 1969.

QUESTIONS AND PROBLEMS

1. Discuss the bond properties of N_2, P_2, As_2, and Bi_2 in terms of their electronic structures.
2. Consider the molecule NF and the ions NF^+ and NF^-. Write the Lewis structure and the molecular-orbital description of the ground state for each species. Determine which of the three species would be paramagnetic and, if so, how many unpaired electrons there would be in each. Predict bond orders for all three species.
3. The ground state of H_2 has the molecular-orbital configuration $(\sigma^b)^2$. In addition to the ground state there are excited states possessing the following configurations:

   Predict which of these states would be highest in energy and which would be lowest. Explain your reasoning. Would you expect the lowest excited state of H_2 to be paramagnetic, or diamagnetic?
4. Discuss the bond properties of Cl_2 and Cl_2^+ using molecular orbital theory.
5. Discuss the electronic structure of the SO molecule in terms of the molecular orbital theory for *AB*-type diatomic molecules. How many unpaired electrons do you predict for the ground state?
6. Describe the electronic structure of LiH in terms of molecular orbital theory. Estimate the partial ionic character in the LiH bond.
7. The OH radical recently has been observed in outer space. Formulate its electronic structure in terms of molecular orbital theory using only the $2p$ oxygen and the $1s$ hydrogen orbitals. What type of molecular orbital contains the unpaired electron? Is this orbital associated with both oxygen and hydrogen atoms, or is it localized on a single atom? Would you expect the lowest electronic transition of OH to occur at lower, or higher, energy than that of OH^-? Briefly explain your answer.
8. Discuss the bond properties of NO, PO, AsO, and SbO. In what important ways does the molecular-orbital description of the ground state of these molecules differ from the Lewis structural model?
9. Discuss the bond properties of the interhalogen diatomic molecules ClF, BrF, BrCl, and ICl. Calculate the partial ionic character in each molecule from its calculated and observed dipole moments. Discuss the results of your calculation in terms of the molecular-orbital model of electronic structure.
10. Formulate the bonding in the hydrogen halide molecules HF, HCl, HBr, and HI in terms of molecular orbital theory. Discuss the bond properties of these molecules.
11. Explain why the dissociation energy is greater for B_2 than for F_2. Can you also rationalize the fact that the F_2 bond distance is *shorter* than the B_2 bond?

12. The mercurous ion is found as the diatomic species Hg_2^{2+}. Ignore the filled *d* orbitals and describe the bonding in this ion in terms of molecular orbital theory. The ion exhibits strong absorption in the ultraviolet region. To what electronic transition can this absorption be attributed?

13. What is the molecular-orbital configuration of the ground state of the CN molecule? How many unpaired electrons does CN have? The molecule has an absorption band in the near-infrared region (at 9000 cm^{-1}), which is due to an electronic transition. Suggest an assignment for this absorption band. The CN^- ion does *not* absorb in the near-infrared region. Is this observation consistent with your transition assignment? If not, it would be a good idea to reconsider!

4

Polyatomic Molecules

The molecular-orbital method that we applied to diatomic molecules provides a logical starting point for understanding polyatomic systems. The most general method for constructing molecular wave functions for polyatomic molecules is to use unhybridized atomic orbitals in linear combinations. Electrons in these molecular orbitals are not localized between two atoms of a polyatomic molecule, rather they are *delocalized* among several atoms. This is conceptually very different from the Lewis picture, in which two electrons between two atoms are equivalent to one chemical bond.

An alternative method for dealing with complex molecules is to use *localized*, two-atom molecular orbitals. For many applications the localized bond theory provides a simple framework for the discussion of ground-state properties, particularly molecular geometry. Thus in this chapter, after comparing the two models in the case of BeH_2, we will emphasize the localized molecular-orbital methods.

4–1 BERYLLIUM HYDRIDE, BeH_2

One of the simplest molecules that we can use to illustrate the delocalized molecular-orbital method is BeH_2, for which we will assume a linear structure. As for a diatomic molecule, we label the molecular axis the Z axis (the H_a–Be–H_b line), as shown in Figure 4–1. The molecular orbitals for BeH_2 are formed by using the $2s$ and $2p$ beryllium valence orbitals and the $1s$ valence orbitals of H_a and H_b. The correct linear combinations for the bonding molecular orbitals are obtained by writing the combinations of $1s_a$ and $1s_b$ that match the algebraic signs on the lobes of the central beryllium atom's $2s$ and $2p_z$ orbitals, respectively. This procedure gives a bonding orbital that concentrates electron density between the nuclei. Since the $2s$ orbital does not change sign over its spherical surface, the combination $(1s_a + 1s_b)$ is appropriate for one bonding molecular orbital (see Figure 4–1). The $2p_z$ orbital has a plus lobe along $+Z$ and a minus lobe along $-Z$. Thus the proper combination of H orbitals for the second bonding molecular orbital is $(1s_a - 1s_b)$, as shown in Figure 4–2.

We can describe the two σ^b molecular orbitals by the following molecular wave functions:

$$\sigma_s^b \propto [2s + \lambda_1(1s_a + 1s_b)]$$
$$\sigma_z^b \propto [2p_z + \lambda_2(1s_a - 1s_b)]$$

The corresponding antibonding molecular orbitals, σ_s^* and σ_z^*, will have nodes between the Be and the two H nuclei. That is, we will combine the beryllium $2s$ orbital with $-(1s_a + 1s_b)$ and the beryllium $2p_z$ orbital with $-(1s_a - 1s_b)$.

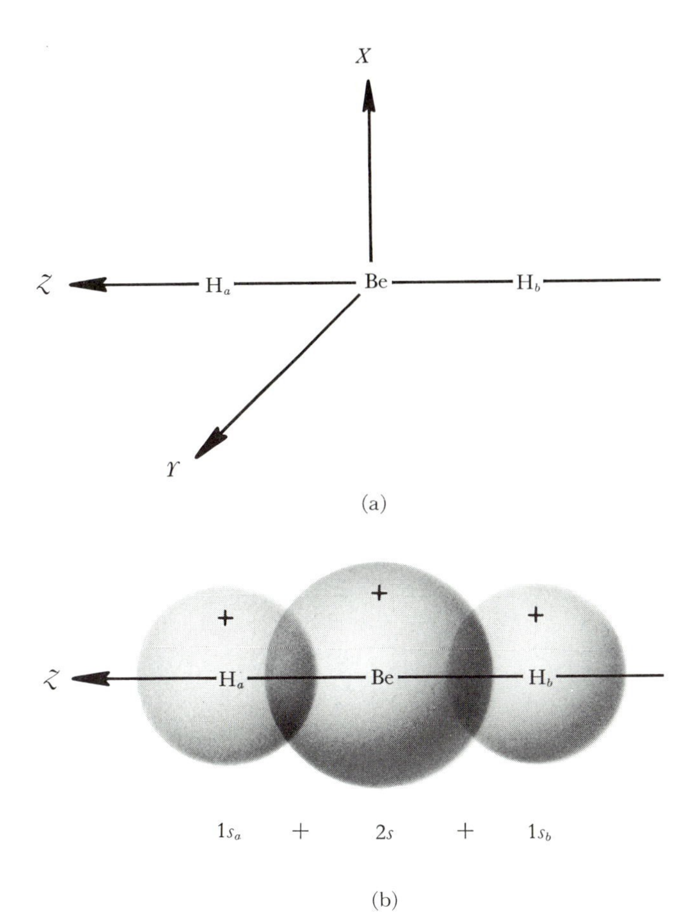

4–1
Linear polyatomic molecule. (a) Coordinate system for BeH_2. (b) Overlap of the hydrogen $1s$ orbitals with the beryllium $2s$ orbital.

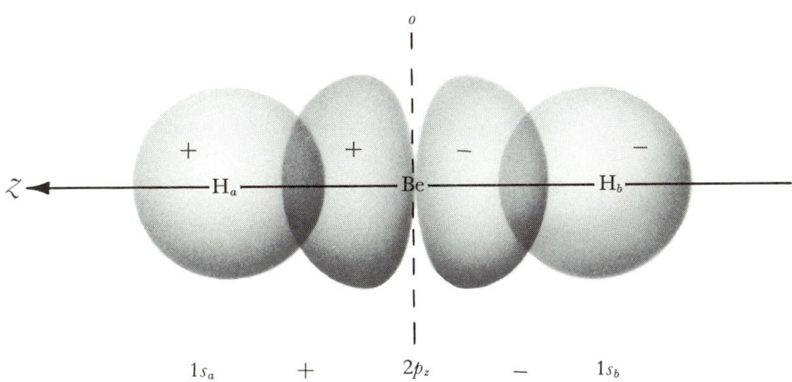

4–2
Overlap of the hydrogen $1s$ orbitals with the beryllium $2p_z$ orbital.

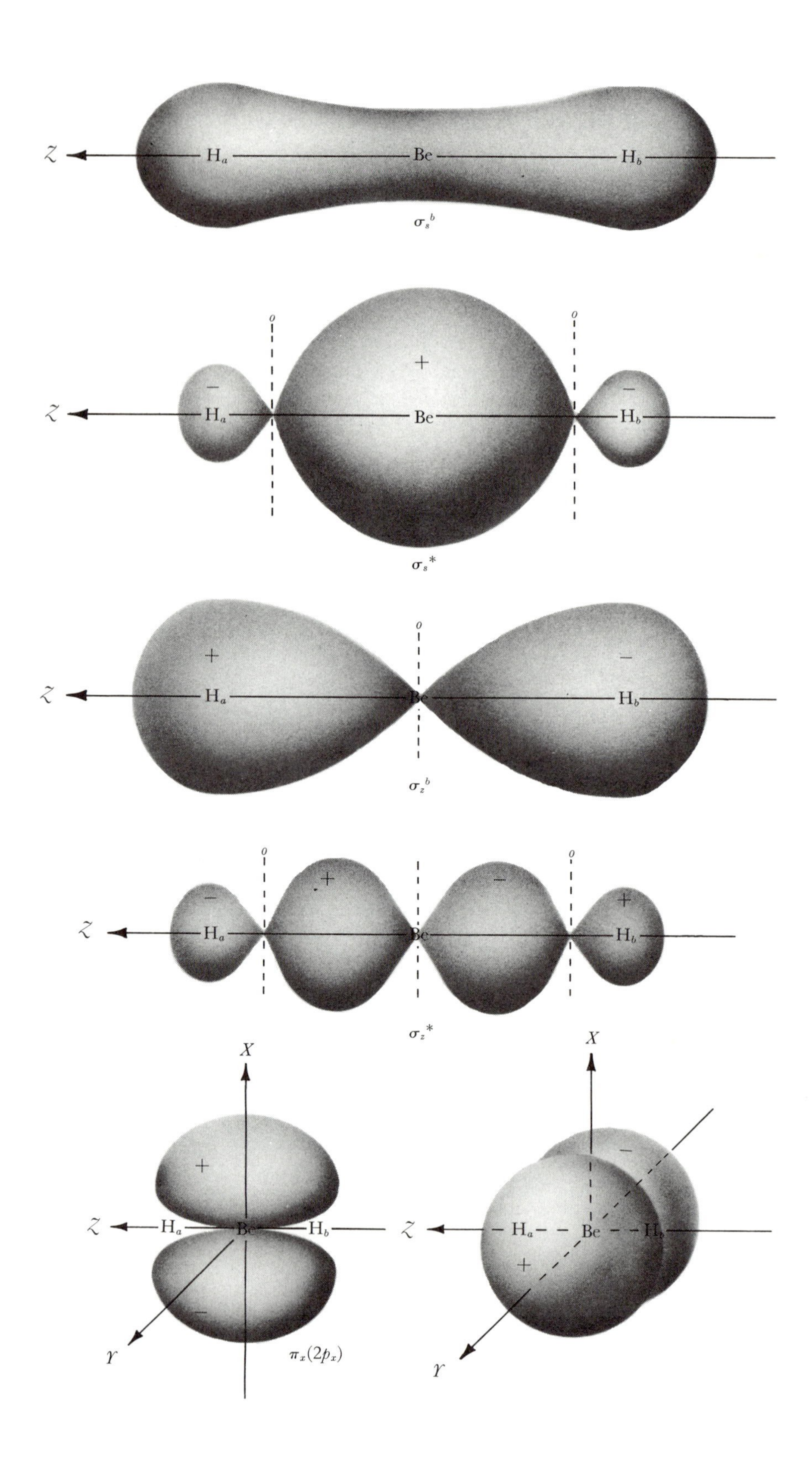

Z
H_a
Be
H_b
$\sigma_s{}^b$
0
0
+
−
−
Z
H_a
Be
H_b
$\sigma_s{}^*$
0
+
−
Z
H_a
Be
H_b
$\sigma_z{}^b$
0
0
−
+
−
+
Z
H_a
Be
H_b
$\sigma_z{}^*$
X
+
Z
H_a
Be
H_b
−
Y
$\pi_x(2p_x)$
X
−
Z
H_a
Be
H_b
+
Y

◀4–3
Spatial representations of the molecular orbitals of BeH_2. The π_x and π_y orbitals shown at the bottom are not used in bonding.

Therefore the two σ^* molecular orbitals are

$$\sigma_s^* \propto [2s - \gamma_1(1s_a + 1s_b)]$$

and

$$\sigma_z^* \propto [2p_z - \gamma_2(1s_a - 1s_b)]$$

Since the beryllium $2s$ and $2p_z$ orbitals are of much higher energy than the hydrogen $1s$ orbitals (H is more electronegative than Be), we can assume confidently that the electrons in the bonding orbitals of BeH_2 spend more time around the hydrogen nuclei. In the antibonding orbitals an electron is forced into the vicinity of the beryllium nucleus. In other words, calculation of the coefficients of the valence orbitals in the molecular-orbital combinations would reveal that λ_1 and λ_2 are greater than γ_1 and γ_2, respectively.

The $2p_x$ and $2p_y$ beryllium orbitals are not used in bonding, because they are π orbitals in a linear molecule and hydrogen has no valence orbitals capable of forming π molecular orbitals. Therefore these orbitals are nonbonding in the BeH_2 molecule. Spatial representations of the BeH_2 molecular orbitals are shown in Figure 4–3.

The molecular-orbital energy-level scheme for BeH_2, shown in Figure 4–4, is constructed as follows. The valence orbitals of the beryllium atom are indicated at the left of the diagram, with the lower-energy $2s$ orbital below the $2p$ orbitals. The $1s$ orbitals of the two hydrogen atoms are placed on the right of the diagram. The $1s$ orbitals of hydrogen are placed lower than either $2s$ or $2p$ orbitals of beryllium because of the difference in electronegativity of the two atoms. The molecular orbitals—bonding, nonbonding, and antibonding—are placed in the middle of the diagram. As usual, the bonding levels are of lower energy than the individual atomic orbitals, and the antibonding levels are of correspondingly higher energy. The energy of the $2p_x$ and $2p_y$ nonbonding Be orbitals does not change in our approximation scheme.

The ground state of BeH_2 is found by placing the valence electrons in the most stable molecular orbitals of Figure 4–4. There are four valence electrons, two from beryllium ($2s^2$) and two from the two hydrogen atoms. Therefore the ground-state electronic configuration is

$$(\sigma_s^b)^2(\sigma_z^b)^2$$

Thus in this description of the electronic structure of BeH_2 the two electron-pair bonds are delocalized over the three atoms.

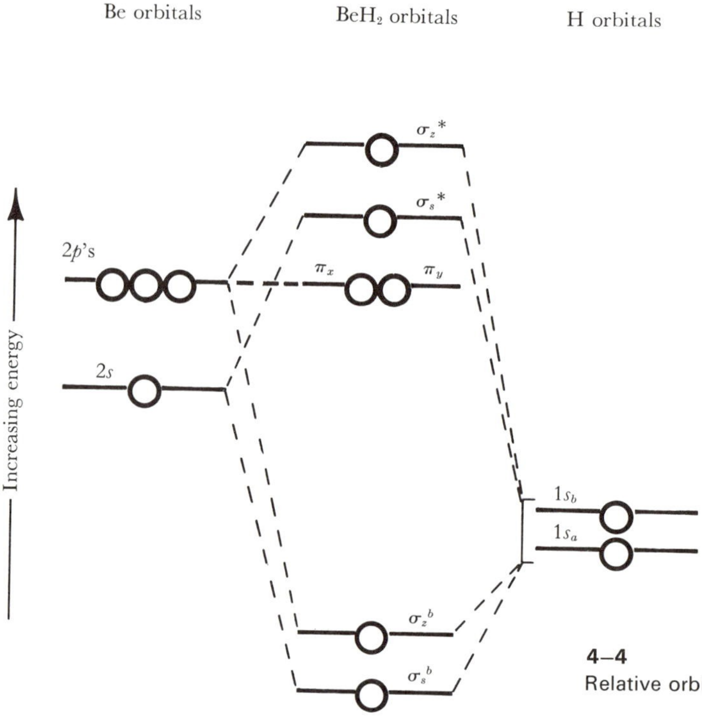

4–4
Relative orbital energies in BeH_2.

4–2 LOCALIZED MOLECULAR ORBITALS FOR BeH_2, BH_3, AND CH_4

A simpler approach to deducing the electronic structures of polyatomic molecules is the localized molecular-orbital method. We will develop this method first for BeH_2 so that we can compare it with the general delocalized molecular orbital theory presented in the preceding section.

Instead of using the pure $2s$ and $2p_z$ beryllium valence orbitals to make two delocalized bonding orbitals, we can construct two equivalent valence orbitals centered on Be, which are directed at H_a and H_b, respectively. This is accomplished by *hybridizing* the $2s$ and $2p$ orbitals to give two equivalent sp hybrid orbitals, as shown in Figure 4–5. One hybrid orbital, sp_a, is directed at H_a and strongly overlaps the $1s_a$ orbital. The other hybrid orbital, sp_b, is directed at H_b and strongly overlaps the $1s_b$ orbital. In this scheme the two bonding molecular orbitals in BeH_2 are formed by making two equivalent linear

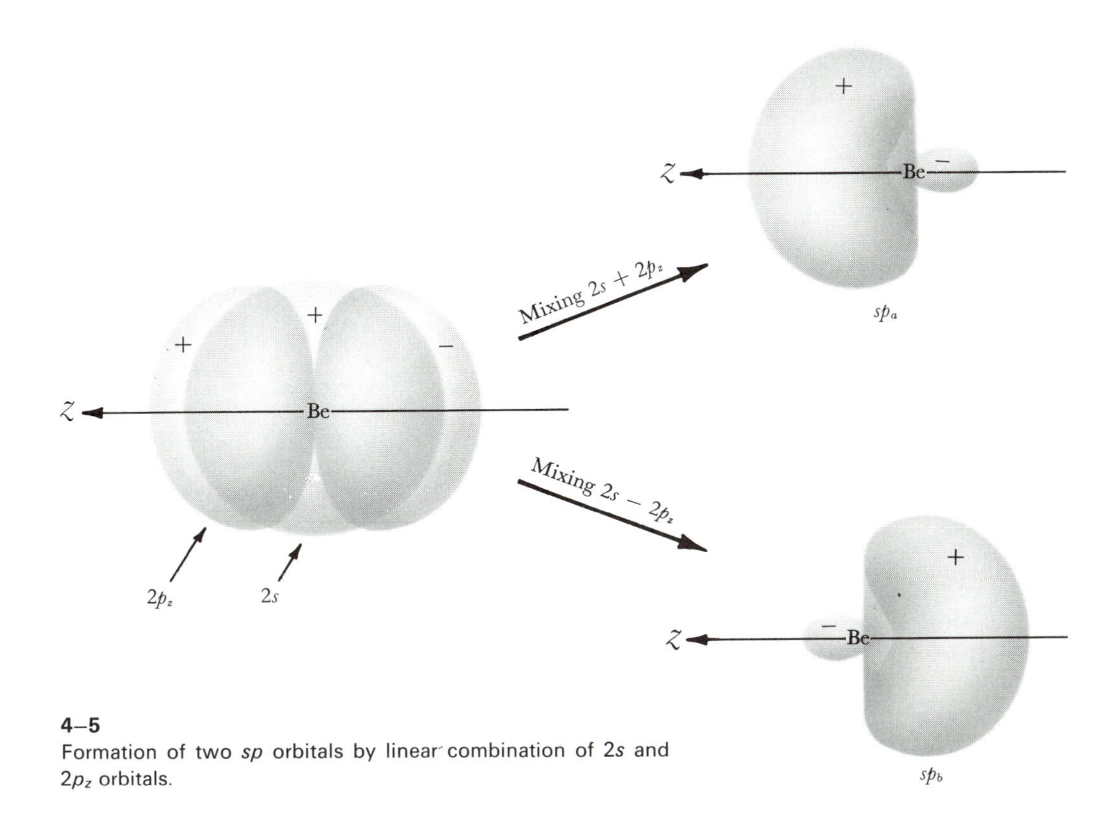

4–5
Formation of two *sp* orbitals by linear combination of $2s$ and $2p_z$ orbitals.

combinations, each of which is localized between two atoms:

$$sp_a + 1s_a$$
$$sp_b + 1s_b$$

The two localized molecular orbitals are shown in Figure 4–6. The four valence electrons participate in two localized electron-pair bonds, analogous to the Lewis structure for BeH_2. Each of the *linear sp* hybrid orbitals has half *s* and half *p* character, and the two *sp* orbitals are sufficient to attach two hydrogen atoms to the central beryllium atom in BeH_2.

Next we will consider the BH_3 molecule (which is observed in a mass spectrometer as a fragment of the B_2H_6 molecule, Section 4–3) in which three hydrogen atoms are bonded to the central boron atom. In the localized molecular-orbital method this is accomplished by hybridizing the $2s$ and $2p$ orbitals of a boron atom to give three equivalent sp^2 hybrid orbitals (Figure

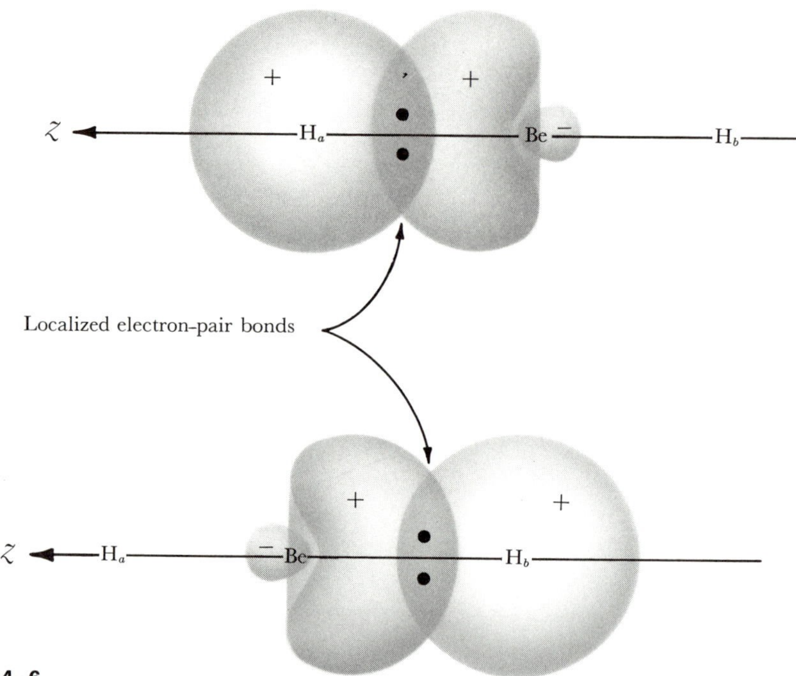

4–6
Localized electron-pair bonds for BeH_2 that are formed from two equivalent *sp* hybrid orbitals centered at the Be nucleus. Each Be *sp* orbital forms a localized bonding molecular orbital with a hydrogen 1*s* orbital.

4–7). Each sp^2 hybrid orbital has one third *s* and two thirds *p* character. Since any two *p* orbitals lie in the same plane and the *s* orbital is nondirectional, the sp^2 hybrid orbitals lie in a plane. The three sp^2 hybrid orbitals form three equivalent, localized bonding orbitals with the three hydrogen 1*s* orbitals. Each of the sp^2–1*s* bonding orbitals is occupied by an electron pair, as illustrated in Figure 4–8. Using hybrid-orbital theory, we would predict the structure of BH_3 to be *trigonal planar*. The angle between the H–B–H internuclear lines, which is called the H–B–H bond angle, is expected to be 120°.

Methane, CH_4, is an example of a molecule with four equivalent atoms attached to a central atom. All the carbon valence orbitals are needed to attach the four hydrogen atoms. Thus by hybridizing the 2*s* and the three 2*p* orbitals

4–7▶
Formation of three equivalent sp^2 hybrid orbitals by linear combination of 2*s* and 2*p* orbitals. The sp^2 hybrid orbitals are trigonal planar.

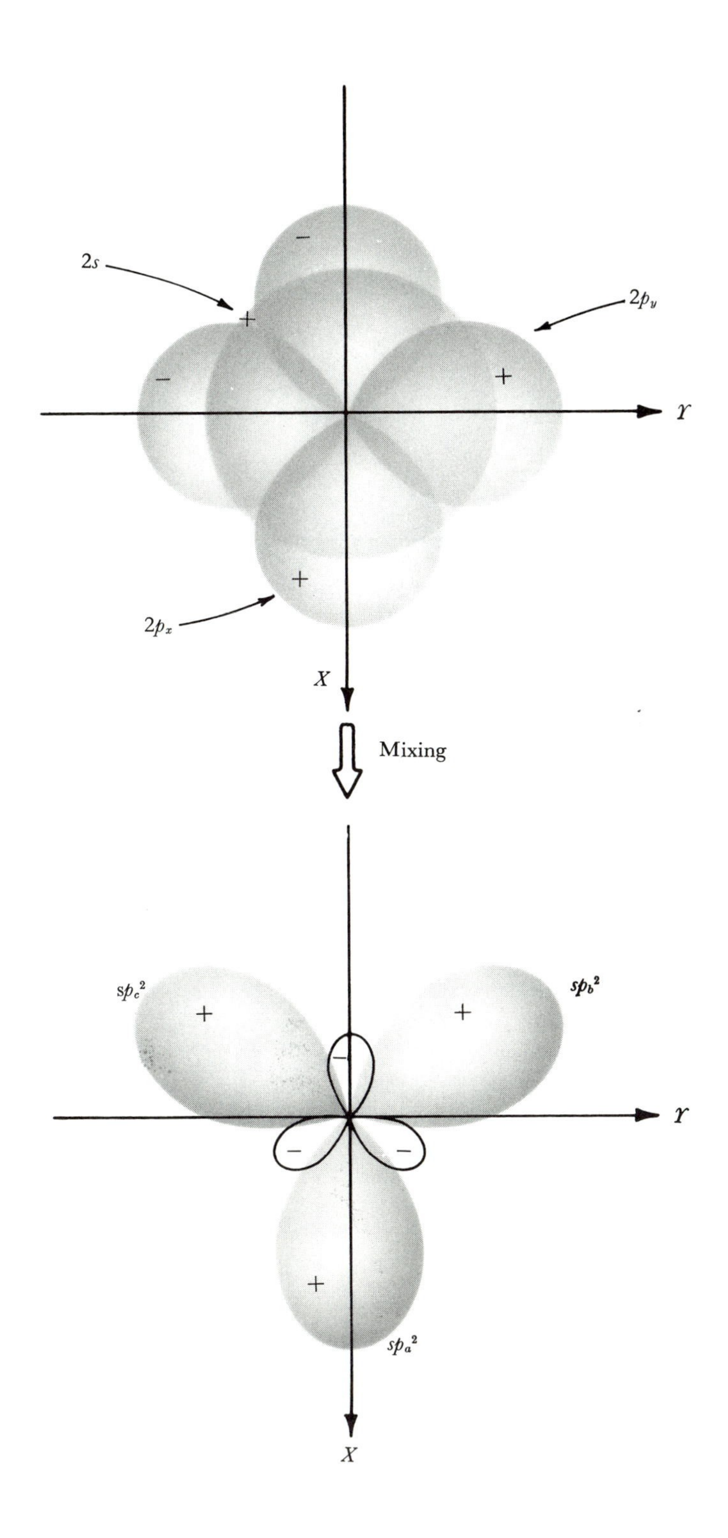

2s
2py
2px
X
Y
Mixing
spc2
spb2
spa2
Y
X

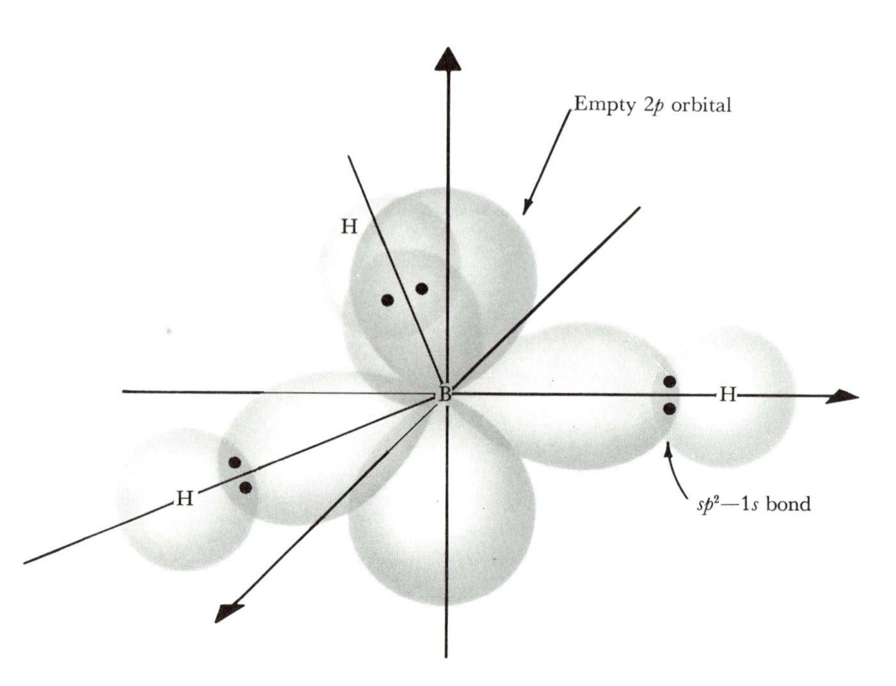

4–8
Localized electron-pair bonds in BH_3.

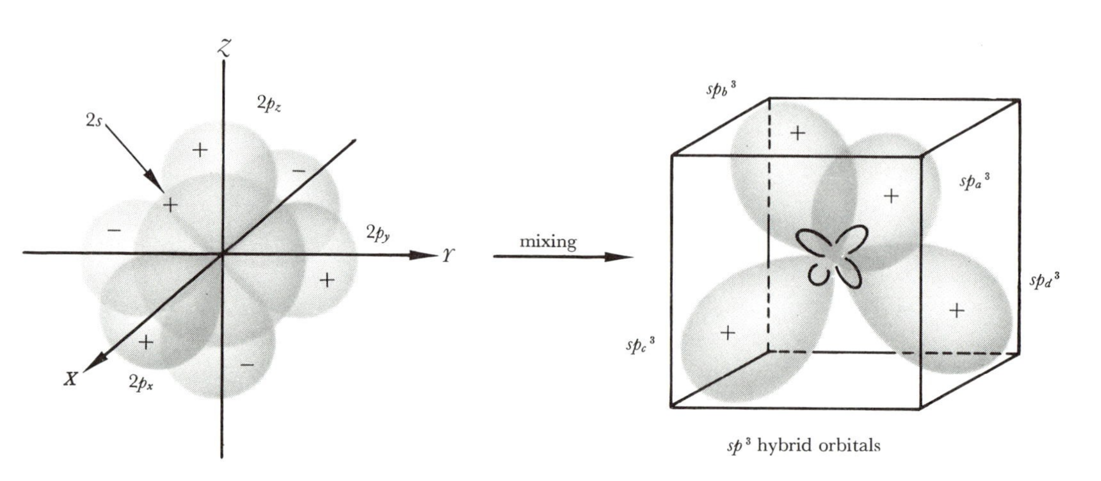

4–9
Formation of four equivalent, tetrahedral sp^3 hybrid orbitals.

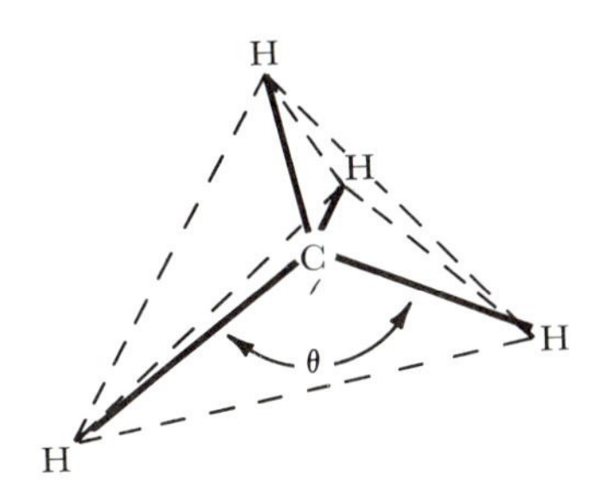

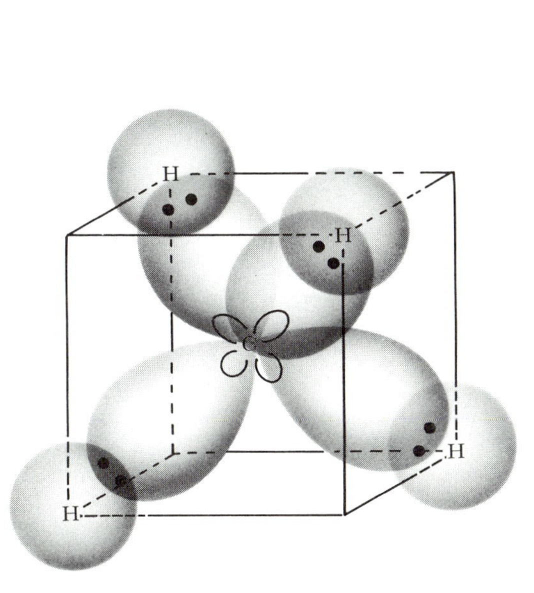

4–10
Localized electron-pair structure for CH_4.

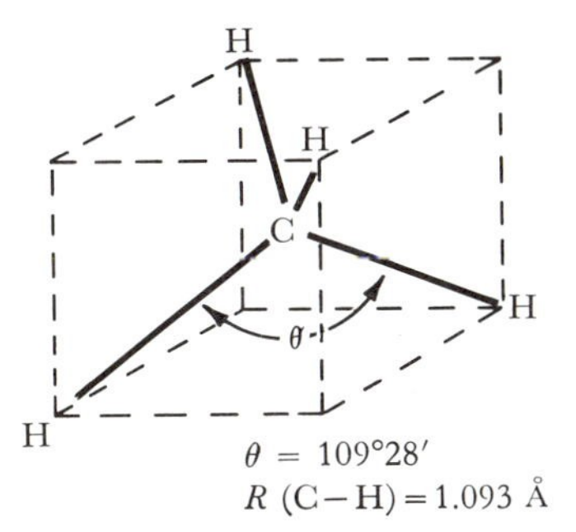

4–11
Tetrahedral molecular structure of CH_4.

we obtain four equivalent sp^3 hybrid orbitals (Figure 4–9). Each sp^3 hybrid orbital has one fourth s and three fourths p character. The four sp^3 orbitals are directed toward the corners of a regular tetrahedron, thus the sp^3 orbitals are called *tetrahedral hybrids*. Four localized bonding orbitals can be made by combining each hydrogen 1s orbital with an sp^3 hybrid orbital. The best overlap between the sp^3 orbitals and the 1s orbitals is obtained by placing the four hydrogen atoms at the corners of a regular tetrahedron, as indicated in Figure 4–10. There are eight valence electrons (four from the carbon atom and one from each of the four hydrogen atoms) to distribute in the four localized bonding orbitals. These eight electrons account for the four equivalent, localized electron-pair bonds shown in Figure 4–10.

The structure of CH_4 has been determined by several experimental methods, including neutron diffraction. All data reveal that the structure of CH_4 is *tetrahedral* (Figure 4–11), which is completely consistent with localized bond-orbital theory. The H–C–H bond angle is 109°28′ and the C–H bond length is 1.093 Å.

A summary of localized bond-orbital theory for BeH_2, BH_3, and CH_4 is given in Table 4–1.

Table 4–1. Localized orbital theory and molecular geometry

Example molecule	Groups attached to central atom	Hybrid orbitals appropriate for central atom	Molecular geometry
BeH_2	2	sp	Linear [angle (H–Be–H) = 180°]
BH_3	3	sp^2	Trigonal planar [angle (H–B–H) = 120°]
CH_4	4	sp^3	Tetrahedral [angle (H–C–H) = 109°28′]

4–3 HYDROGEN IN BRIDGE BONDS

Of the three molecules discussed in the preceding section, only CH_4 has a closed valence-shell bonding configuration. At normal temperatures and pressures both BeH_2 and BH_3 use their empty valence orbitals to form larger molecular aggregates. Beryllium hydride is a solid in which hydrogen atoms share electrons with adjacent beryllium atoms in *"bridge" bonds*, which may be represented as

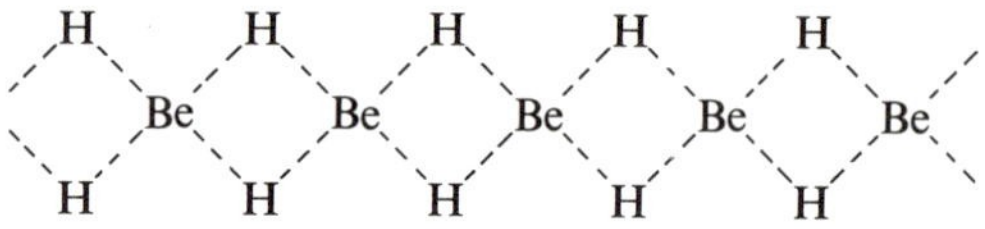

H H H H H
Be Be Be Be Be
H H H H H

In a sense each Be shares eight electrons in the solid, thereby achieving a closed valence shell.

Under ordinary conditions the compound of empirical formula BH_3 has the molecular formula B_2H_6 and is called *diborane*. The experimentally determined structure of B_2H_6 reveals two types of hydrogen atoms, as shown in Figure 4–12. Two BH_2 units are held together through two B–H–B bridge, or *three-center*, bonds. In the B_2H_6 structure the regular (or terminal) B–H bond length is shorter than the B---H distance in the bridge bonds.

One way to formulate the electronic structure of B_2H_6 employing localized molecular orbitals is shown in Figure 4–13. Each boron atom uses two sp^3 hybrid orbitals to attach the two terminal hydrogen atoms. Each of the remaining two sp^3 orbitals forms a three-center bonding orbital with a hydrogen $1s$ orbital and an sp^3 orbital on the other boron atom. In this model (Figure 4–12) the bridging hydrogen atoms are positioned above and below the plane of the connected BH_2 fragments, as is observed experimentally.

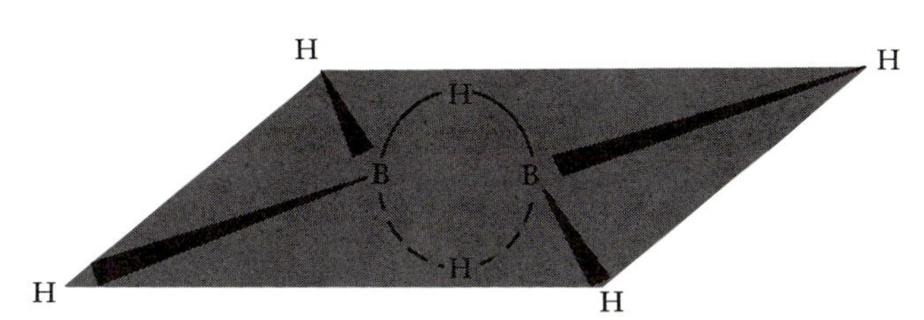

4–12
Bridging bonds in diborane. The arc from B through H to B represents a three-center electron-pair bond—one bonding molecular orbital spread over three nuclei with a capacity for two electrons. The B–H distance in the bridge bonds is 1.334 ± 0.027 Å, as compared to the terminal B–H bond length of 1.187 ± 0.030 Å.

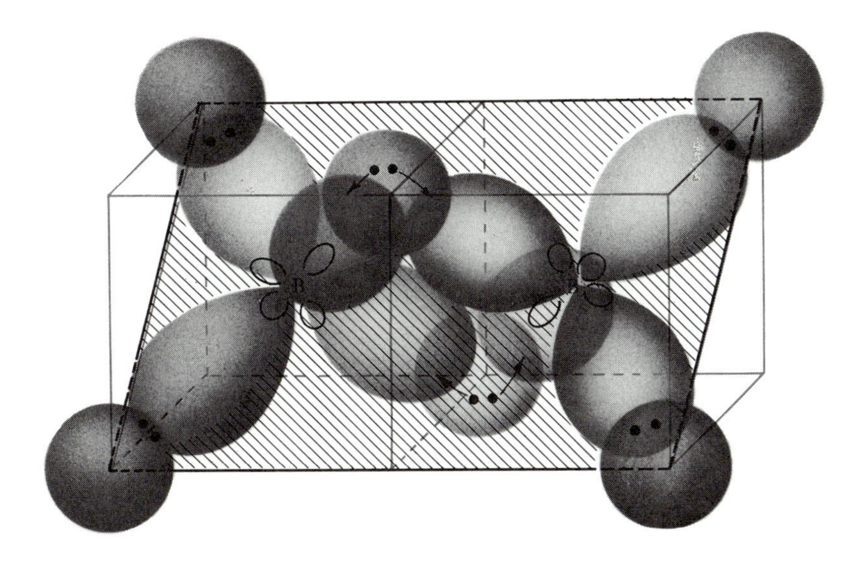

4–13
A localized-orbital model of the three-center electron-pair bonds in diborane, B_2H_6. The 1*s* orbitals of bridging hydrogen atoms overlap with the sp^3 hybrid orbitals on each B atom.

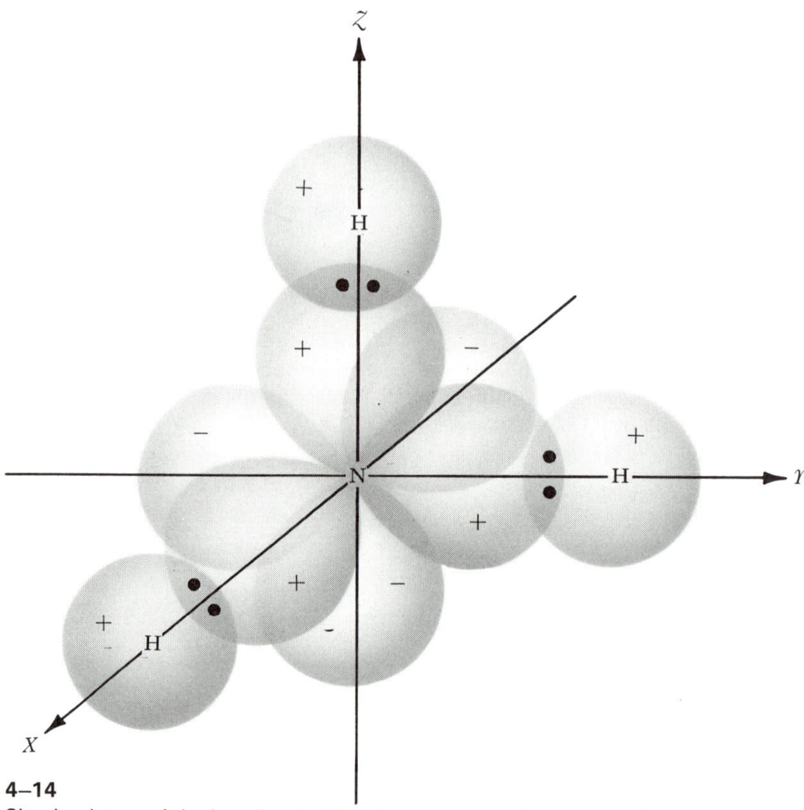

4–14
Simple picture of the bonding in NH_3 that uses only the nitrogen 2*p* orbitals.

4–4 LOCALIZED MOLECULAR ORBITAL THEORY FOR MOLECULES WITH LONE ELECTRON PAIRS—NH_3 AND H_2O

The NH_3 molecule is similar to CH_4 in that it has four valence electron pairs associated with the central atom:

```
     H
     |
  H—N̈—H
```

However, in NH_3 not all of these electron pairs are equivalent. As we can see from its Lewis structure, NH_3 has three N–H single bonds and one lone electron pair. It also is known that the three hydrogen atoms in NH_3 are equivalent. One simple formulation of the bonding in NH_3 involves three localized electron-pair bonds between the nitrogen 2*p* and the hydrogen 1*s* orbitals (Figure 4–14). In this model the lone electron pair is in the nitrogen 2*s* orbital.

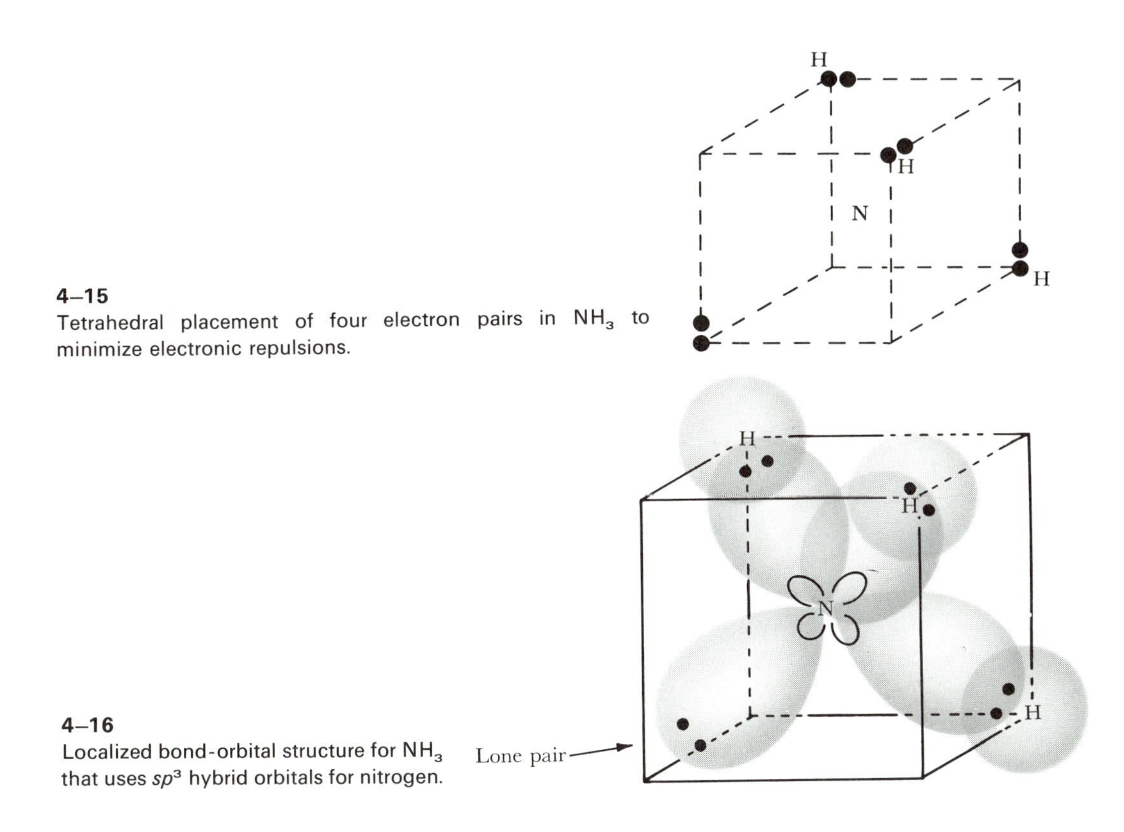

4–15
Tetrahedral placement of four electron pairs in NH_3 to minimize electronic repulsions.

4–16
Localized bond-orbital structure for NH_3 that uses sp^3 hybrid orbitals for nitrogen.

Another simple bonding scheme for NH_3 emphasizes minimizing the repulsions of the four valence electron pairs; the most stable arrangement for four electron pairs is tetrahedral (Figure 4–15). To accommodate the electrons in the tetrahedral arrangement, we use four equivalent sp^3 hybrid orbitals on the nitrogen atom. In this model the nitrogen $2s$ orbital is involved in the N–H bonds. Each of the three N–H bonding orbitals is constructed from one nitrogen sp^3 hybrid orbital and one hydrogen $1s$ orbital. The lone electron pair is assigned to the remaining sp^3 hybrid orbital. This model of the electronic structure of NH_3 is shown in Figure 4–16.

The two models of the electronic structure of NH_3 allow us to predict different H–N–H bond angles, although both predictions give the same general molecular shape. By the molecular shape we mean the positions of the atoms, which we can determine experimentally, but not the placement of lone-pair electrons, which only can be inferred. From both the $(2p + 1s)$ and $(sp^3 + 1s)$

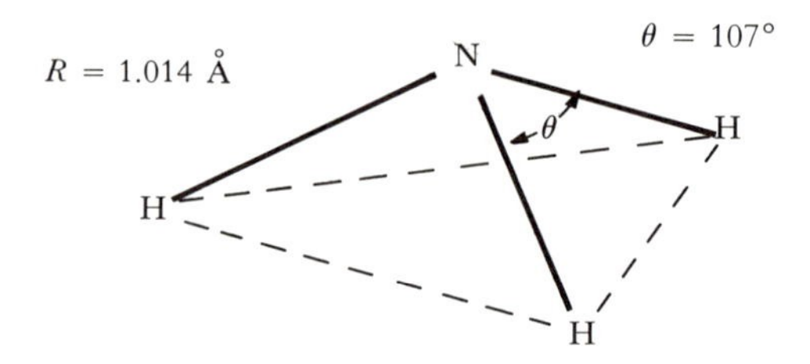

4–17
A view of the trigonal pyramidal molecular structure of NH_3.

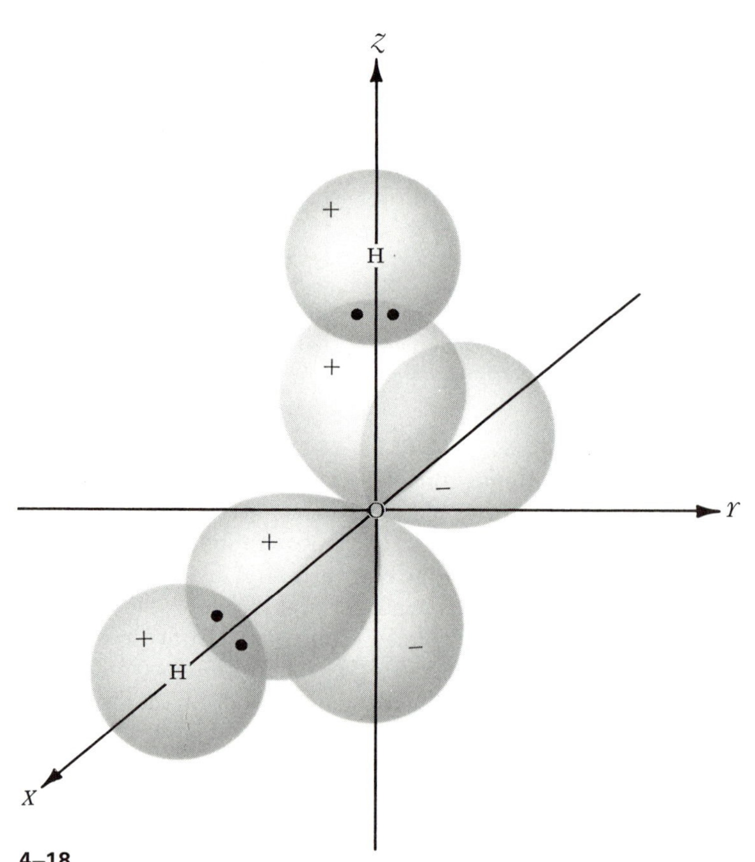

4–18
Simple picture of the bonding in H_2O. Only the oxygen $2p$ orbitals are used, thus an H–O–H bond angle of 90° is predicted.

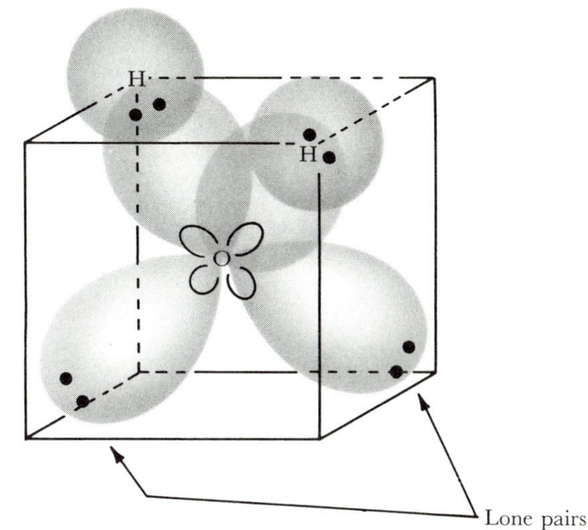

4–19
Localized-bond structure for H_2O with sp^3 orbitals for oxygen. Using this model we predict an H–O–H bond angle of 109° 28′.

bonding schemes we predict that the molecular shape of NH_3 is *trigonal pyramidal*. However, from the ($2p + 1s$) scheme, we predict an H–N–H bond angle of 90° (the angle between the *p* orbitals), whereas from the ($sp^3 + 1s$) scheme, we predict the tetrahedral angle of 109° 28′ (the angle between the sp^3 hybrid orbitals).

The experimentally determined molecular structure of NH_3 is shown in Figure 4–17. Ammonia is trigonal pyramidal and the observed H–N–H bond angle is 107°, which is much closer to the tetrahedral angle predicted from the ($sp^3 + 1s$) model. Thus we conclude that repulsions of valence electron pairs play an important role in determining molecular geometry. The N–H bond length is 1.014 Å, which is slightly shorter than the C–H bond length of 1.093 Å in CH_4. This shorter N–H bond length is consistent with the fact that the nitrogen atom is smaller than the carbon atom.

A similar approach can be taken to deduce the electronic structure of the water molecule. The Lewis structure of H_2O is

$$\begin{array}{c} \mathrm{H{-}\ddot{O}:} \\ \quad\ | \\ \quad\ \mathrm{H} \end{array}$$

with two H–O single bonds and two lone electron pairs. Since atomic oxygen has the valence electronic configuration $2s^22p^4$, we can construct two localized electron-pair bonds, using ($2p + 1s$) bonding orbitals (Figure 4–18). In this scheme the lone electron pairs are in the $(2s)^2$ and $(2p_y)^2$ orbitals. If we consider the repulsions of the four electron pairs to be of primary importance, the 2*s* and the three 2*p* orbitals of oxygen must be hybridized to make four sp^3 orbitals. Two ($sp^3 + 1s$) bonding orbitals are formed between the oxygen atom and the two hydrogen atoms. The two lone electron pairs are equivalent and occupy the remaining two sp^3 orbitals. This model of the electronic structure of H_2O is shown in Figure 4–19.

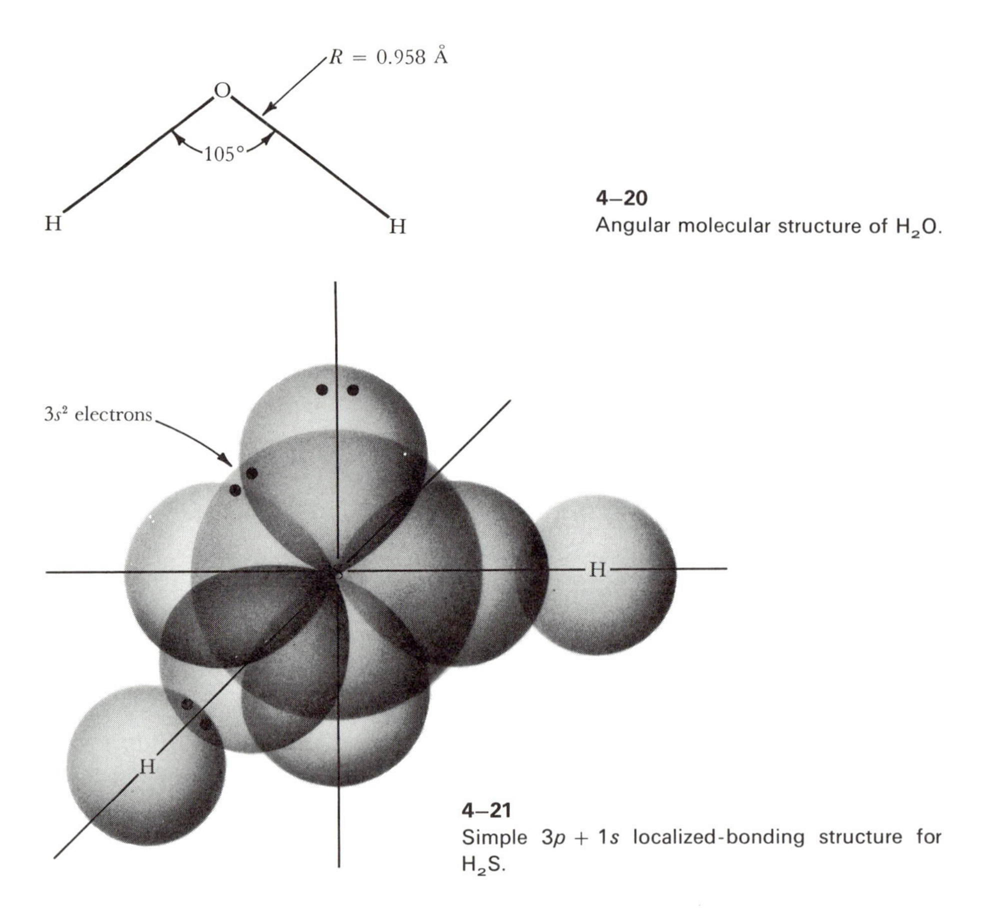

4–20
Angular molecular structure of H_2O.

4–21
Simple $3p + 1s$ localized-bonding structure for H_2S.

From both electronic structural descriptions we predict correctly that the shape of the H_2O molecule is angular. With the $(2p + 1s)$ bonding scheme we predict an H–O–H bond angle of 90°, whereas with the $(sp^3 + 1s)$ bonding scheme the angle would be 109° 28′. The actual molecular geometry of H_2O is illustrated in Figure 4–20. The observed H–O–H bond angle of 105° again is much closer to the angle predicted from the $(sp^3 + 1s)$ model, which minimizes the repulsions of the four electron pairs.

As the central atom in a molecule becomes larger the electrons in valence orbitals, on the average, are farther from each other. Consequently interelectronic repulsions play a proportionately smaller role in determining molecular shapes. For example, a sulfur atom is effectively larger than atomic oxygen, and from atomic spectra it is known that interelectronic repulsions in the sulfur valence orbitals are substantially smaller than in the oxygen valence orbitals. This probably is the reason why the H–S–H bond angle in hydrogen sulfide (H_2S) is 92°, which is much closer to the bond angle that is predicted from the $(3p + 1s)$ bonding model (Figure 4–21). Apparently the repulsion of the two bonding electron pairs in H_2S is much less than the repulsion of the two bonding pairs in H_2O.

4–5 THE VALENCE-SHELL ELECTRON-PAIR REPULSION (VSEPR) METHOD AND MOLECULAR GEOMETRY

Our discussion of the bond angles in H_2O and NH_3 established the principle that bonding electron pairs and lone electron pairs will adopt a spatial arrangement that minimizes electron-pair repulsion. Application of this principle to predict molecular shapes commonly is called the *valence-shell electron-pair repulsion* (VSEPR) method.

The VSEPR method logically accompanies the localized molecular orbital theory. To apply the VSEPR method we simply count the number of lone electron pairs and the number of bonding σ electron pairs around the central atom in a polyatomic molecule. We will call the total number of electron pairs the *steric number*, SN. The sp, sp^2, and sp^3 hybrid orbitals are appropriate for molecules in which SN = 2, 3, and 4, respectively. For SN = 5, we must combine one s, three p, and one d valence orbitals to give five *trigonal bipyramidal* (sp^3d) hybrid orbitals. A set of six equivalent, octahedrally directed sp^3d^2 orbitals is necessary for SN = 6. Examples of predicted molecular shapes from SN = 2 through SN = 6 are shown in Figure 4–22.

The placement of electron pairs for SN = 5 shown in Figure 4–22 deserves special comment. The spatial arrangement for five electron pairs is a trigonal bypyramid (sp^3d hybrids), in which there are three equatorial and two axial positions. The shape of a molecule such as PF_5 must be a trigonal bipyramid because there are no lone pairs. But where should we place the lone electron pair in SF_4? According to the VSEPR method, the most prohibitive repulsion is (lone pair)-(lone pair), followed in order by (lone pair)-(bonded pair) and (bonded pair)-(bonded pair). Therefore in SF_4 the "worst" repulsion is (lone pair)-(bonded pair), because there is only one lone pair. If we place the lone pair in one of the equatorial orbitals, it will repel only two bonded pairs at a 90° angle, whereas axial placement results in three 90°-angle repulsions:

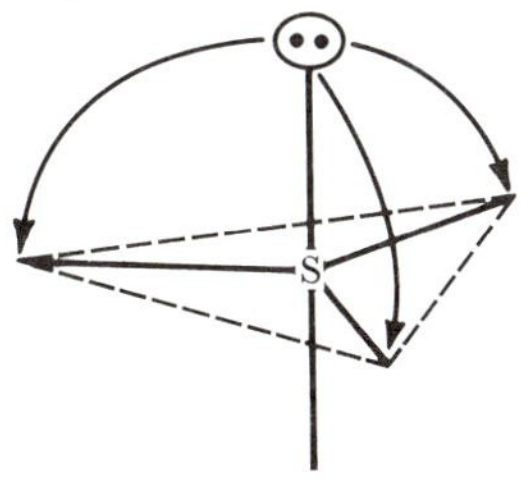

three 90° interactions
(axial placement of lone pair)

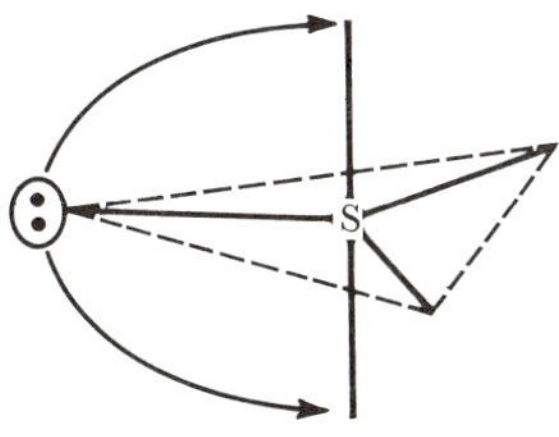

two 90° interactions
(equatorial placement of lone pair)

SN	Number of lone pairs	Orbital hybridization	Molecular shape	Example
2	0	*sp*	 linear	BeH_2, CO_2
3	0	sp^2	 trigonal planar	SO_3, BF_3
3	1		 angular	SO_2, O_3
4	0	sp^3	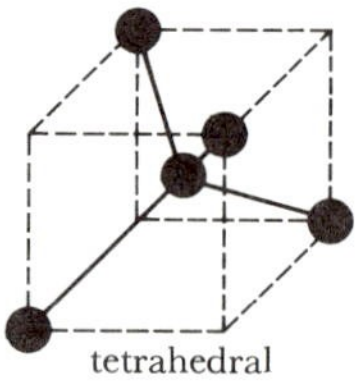 tetrahedral	CH_4, CF_4, SO_4^{2-}
4	1		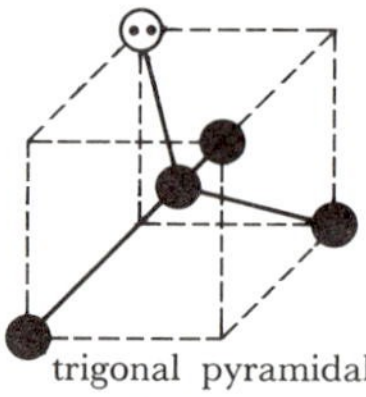 trigonal pyramidal	NH_3, PF_3, $AsCl_3$
4	2		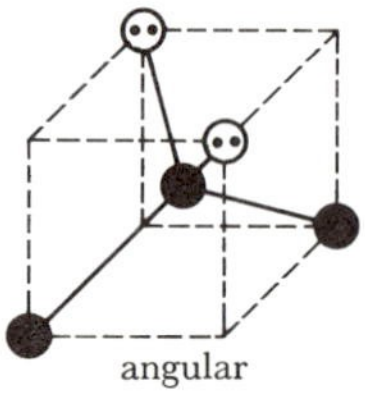 angular	H_2O, H_2S, SF_2
5	0	sp^3d	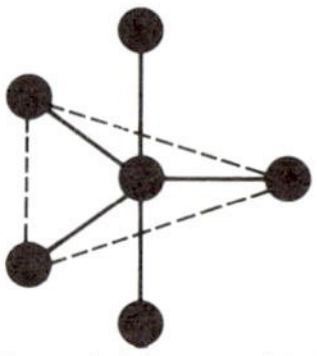 trigonal bipyramidal	PF_5, PCl_5, AsF_5

4–22
Molecular shapes predicted from the VSEPR method.

SN	Number of lone pairs	Orbital hybridization	Molecular shape	Example
5	1		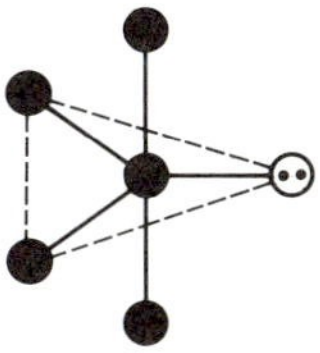 sawhorse	SF_4
5	2		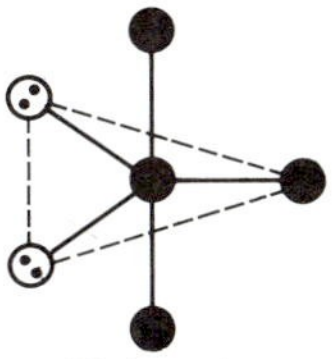 T-shaped	ClF_3
5	3		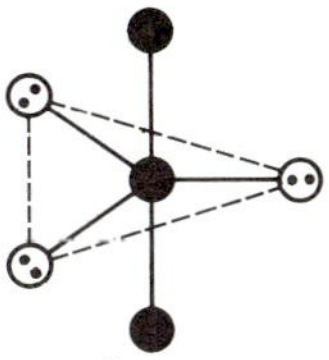 linear	XeF_2, I_3^-, IF_2
6	0	sp^3d^2	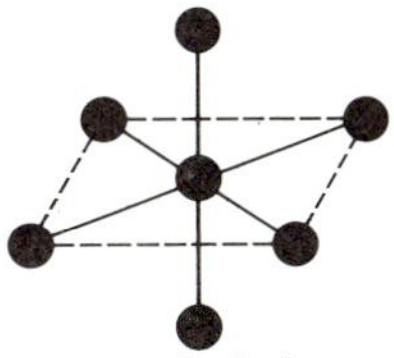 octahedral	SF_6, PF_6^-, SiF_6^{2-}
6	1		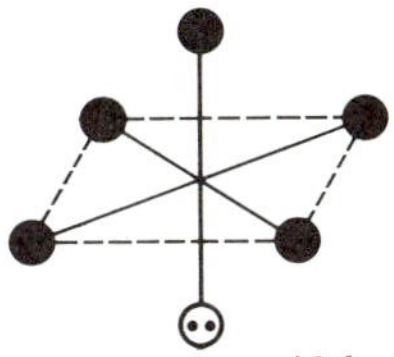 square pyramidal	IF_5, BrF_5
6	2		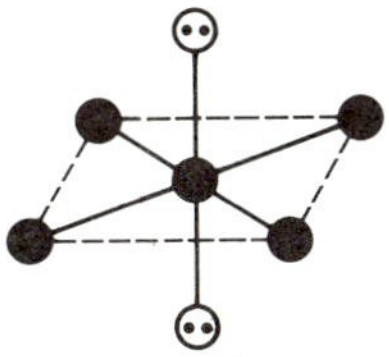 square planar	XeF_4, IF_4^-

4–23 ▶
Contributions to the 1.844 D dipole moment of H_2O.

The 90° interactions are much larger than 120° or 180° interactions, because electron-pair repulsions fall off very rapidly as the distance between pairs increases (see Section 6–2). Therefore the VSEPR choice is equatorial placement of the lone electron pair, which results in a smaller number of 90° interactions. Similar reasoning leads to the placement of the second (e.g., ClF_3) and third (e.g., I_3^-) lone pairs into equatorial orbitals and to the predicted shapes shown in Figure 4–22.

The VSEPR method is easy to use and gives the correct molecular shape for a remarkably large number of molecules. For example, all of the predicted shapes in Figure 4–22 are in agreement with experimentally determined molecular structures.

4–6 POLAR AND NONPOLAR POLYATOMIC MOLECULES

An example of a *polar* polyatomic molecule is H_2O. Because the oxygen valence orbitals are of lower energy than the hydrogen 1*s* orbitals (oxygen is more electronegative than hydrogen), the electron pairs in the two O–H bonds are pulled more toward the oxygen atom. In addition, the oxygen atom has two lone pairs of electrons. The result is a separation of charge in the H_2O molecule, in which the oxygen atom is relatively negative and the hydrogen atoms are relatively positive:

$\delta-$:O: ; $\delta+$ H ; H $\delta+$

Because of the angular shape of H_2O, the H–O bonds and lone-pair contributions combine, as shown in Figure 4–23, to give the dipole moment of 1.844 D.

A nonpolar molecule has zero (or nearly zero) dipole moment. A molecule such as H_2, in which the bonding electron pair is shared equally by both atoms, has zero dipole moment and is *nonpolar*. Molecules in which valence electrons are shared unequally also may have zero dipole moments if the shapes of the molecules are symmetrical. An example is CCl_4. The molecular and electronic structures of CCl_4 are shown in Figure 4–24. Since chlorine is more electronegative than carbon, the bonding electron pairs are pulled toward the chlorine atoms. Thus each C–Cl unit has a small bond dipole moment. The bond dipoles can be resolved into equal and opposite CCl_2 dipoles, as shown in Figure 4–25. The symmetrical (tetrahedral) molecular shape of CCl_4 results in a zero dipole moment, thus CCl_4 is nonpolar.

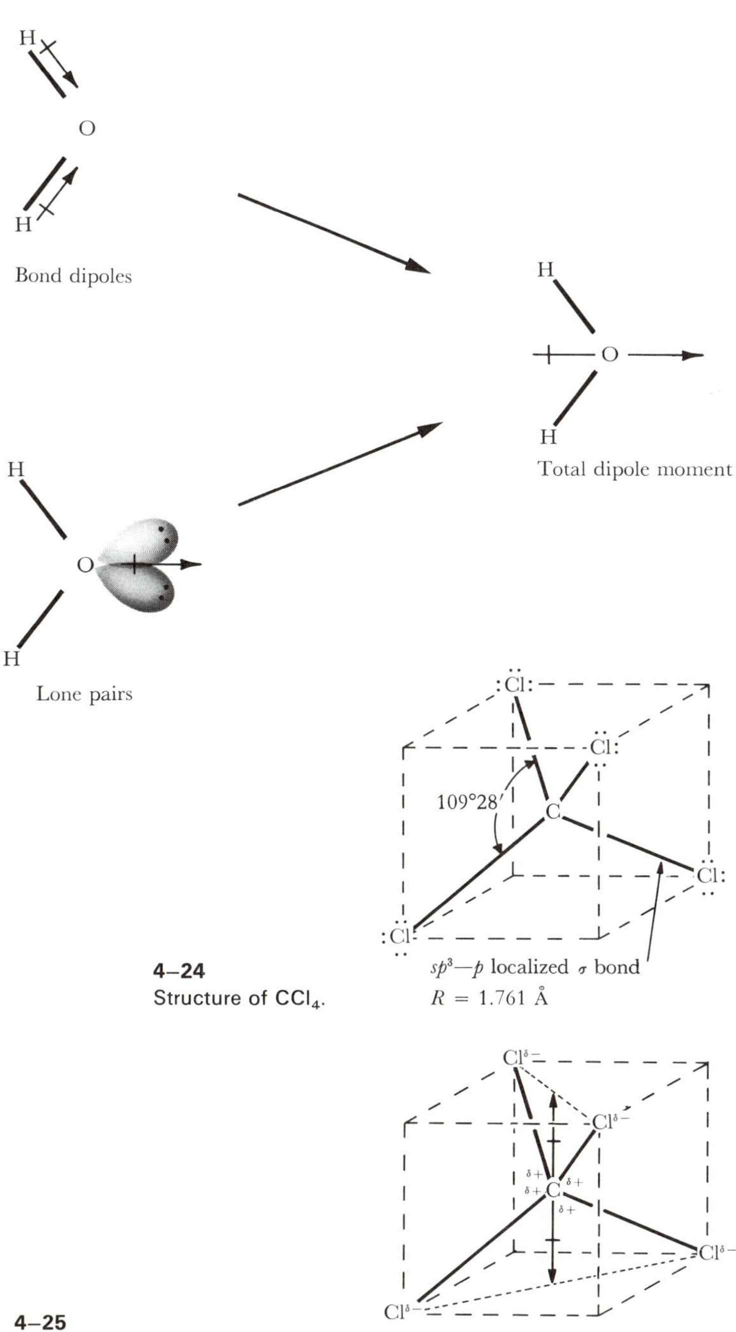

4–24
Structure of CCl_4.

4–25
Canceling bond dipoles in CCl_4.

Table 4–2. Molecular shapes and dipole moments of selected polyatomic molecules

Molecule	Shape	Dipole moment, D	Classification
CS_2	Linear	0	Nonpolar
HCN	Linear	2.98	Polar
COS	Linear	0.712	Polar
BF_3	Trigonal planar	~0	Nonpolar
SO_3	Trigonal planar	~0	Nonpolar
CF_4	Tetrahedral	~0	Nonpolar
CCl_4	Tetrahedral	~0	Nonpolar
NH_3	Trigonal pyramidal	1.47	Polar
PF_3	Trigonal pyramidal	1.03	Polar
AsF_3	Trigonal pyramidal	2.59	Polar
H_2O	Angular	1.85	Polar
H_2S	Angular	0.97	Polar
SO_2	Angular	1.63	Polar
NO_2	Angular	0.316	Polar
O_3	Angular	0.53	Polar

In summary, polar molecules have bond dipoles that add to give a resultant nonzero dipole moment. Nonpolar molecules have either pure covalent bonds (equal sharing) or bond dipoles that cancel due to a symmetrical molecular shape. Table 4–2 gives the molecular shapes and the polarities of several representative polyatomic molecules.

4–7 SINGLE AND MULTIPLE BONDS IN CARBON COMPOUNDS

Carbon atoms have a remarkable ability to form bonds with hydrogen atoms and other carbon atoms. Because a carbon atom has one $2s$ and three $2p$ valence orbitals, the structure around a carbon atom for full σ bonding is tetrahedral (sp^3). If one hydrogen atom in CH_4 is replaced with a CH_3 group, the C_2H_6 (ethane) molecule is obtained. The C_2H_6 molecule contains one C–C bond, and the structure around each carbon atom is tetrahedral (sp^3), as shown in Figure 4–26. By continually replacing hydrogen atoms with CH_3 groups, we obtain the many hydrocarbons with the full sp^3 σ-bonding structure at each carbon atom. Such hydrocarbons are said to be *saturated* because each carbon atom uses its valence orbitals to attach the maximum number (4) of atoms through σ bonds.

Ethylene

In *unsaturated* organic molecules a carbon atom uses only three or two of its four valence orbitals for σ bonding. This leaves one or two $2p$ orbitals available

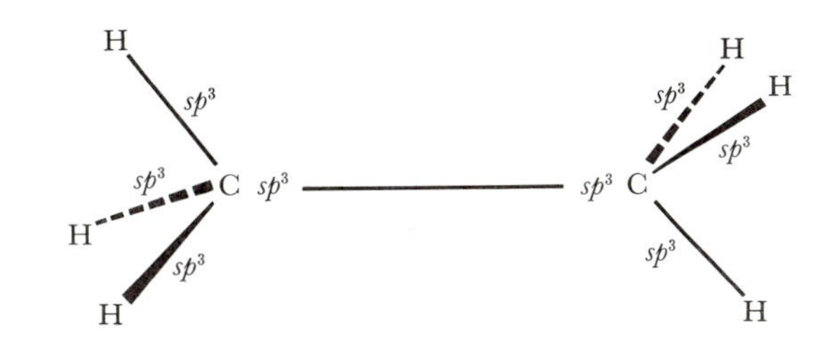

4–26
Localized-bond structure for C_2H_6.

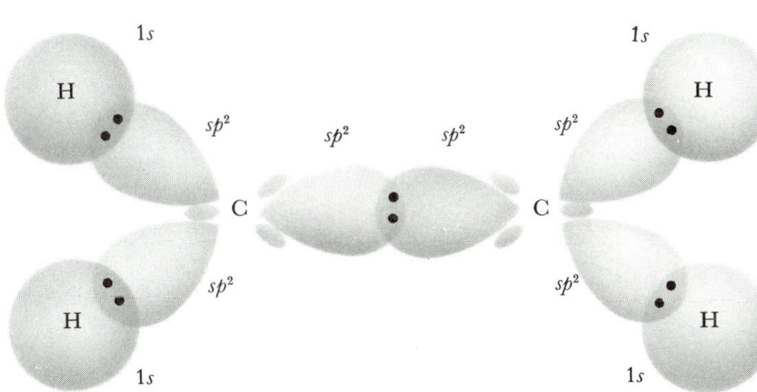

4–27
The σ-bond structure for C_2H_4.

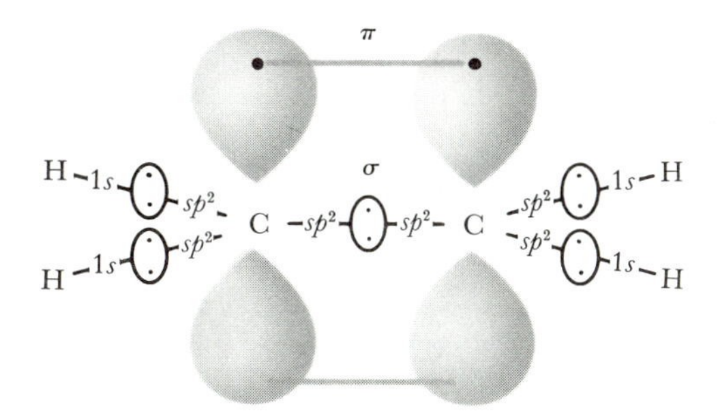

4–28
Representation of the bonding in C_2H_4, using localized molecular orbitals with $2p$–$2p$ overlap to form a π bond.

for π bonding. For example, in ethylene, C_2H_4, there are three atoms around each carbon atom. One way to make localized molecular orbitals for C_2H_4 is first to use three sp^2 hybrid orbitals for each carbon atom to form five σ bonds, four C–H bonds and one C–C bond (Figure 4–27). The σ bonds account for 10 of the 12 valence electrons in C_2H_4. In this scheme each carbon atom has one valence p orbital not involved in σ bonding that is perpendicular to the set of sp^2 orbitals. Thus the two $2p$ carbon orbitals in ethylene can overlap to form a π bonding molecular orbital. The $2p - 2p$ overlap is largest when the two sp^2 orbital sets are oriented in the same plane, as shown in Figure 4–28.

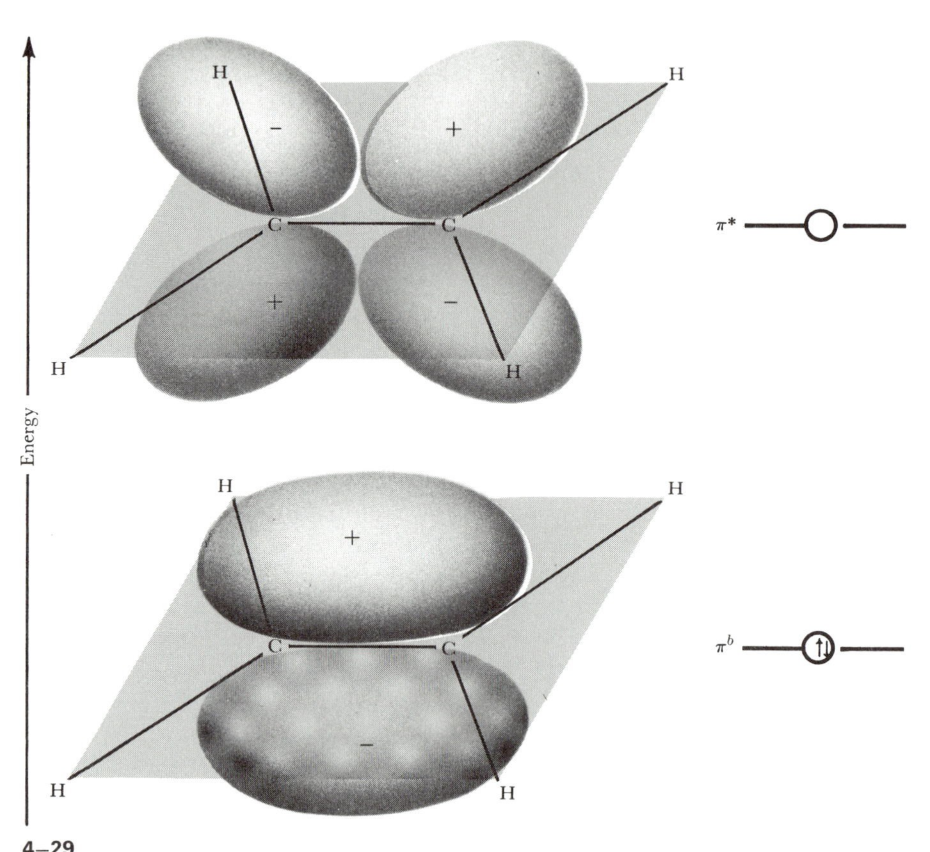

4–29
The bonding and antibonding π molecular orbitals in C_2H_4. The π electronic structure of the ground state is $(\pi^b)^2$.

The remaining valence electron pair in C_2H_4 occupies the π bonding orbital. From the electronic structural model shown in Figure 4–28 we predict that the C_2H_4 molecule will have a planar structure, with H–C–H and H–C–C bond angles of 120°. From experimental data we know that the molecule is planar, and that the H–C–H bond angle is 117° and the H–C–C bond angle is 121° 31′. Thus the molecular structure of C_2H_4 is in close agreement with the predictions from the electronic structural model.

Unsaturated hydrocarbons such as ethylene absorb light at longer wavelengths (lower energies) than do saturated hydrocarbons. For example, ethylene has an ultraviolet absorption peak at 1710 Å (58,500 cm^{-1}), whereas ethane does not begin to absorb strongly before 1600 Å. This fact suggests that the separation between σ bonding and σ antibonding orbitals in hydrocarbons is larger than the separation between π bonding and π antibonding orbitals.

For this reason it is common to ignore the σ^* levels, as we have done, but to include both π^b and π^* orbitals in a molecular-orbital formulation of unsaturated hydrocarbons. Spatial representations and relative energies of the π^b and π^* orbitals of C_2H_4 are shown in Figure 4–29. On the basis of the π-orbital electronic structure of C_2H_4 we assign the absorption peak at 58,500 cm^{-1} to the electronic transition $\pi^b \rightarrow \pi^*$.

If the π-electron system of an unsaturated hydrocarbon is more extensive than ethylene, the energy separation between the highest occupied π^b orbital and the lowest unoccupied π^* level becomes smaller, and energy absorption occurs at longer wavelengths. Such extensive π-electron systems are found in *conjugated polyenes*, compounds in which conventional structural formulas show alternate single and double bonds:

$$—\overset{|}{C}=\underset{|}{C}—\overset{|}{C}=\underset{|}{C}—\overset{|}{C}=\underset{|}{C}—$$

conjugated polyene skeleton

Polyenes having ten or more conjugated double bonds absorb visible light, hence they are colored. The pigments responsible for light perception in the human eye contain long, conjugated polyene chains, as do some vegetable pigments such as carotene, the colored substance in carrots.

Saturated and unsaturated carbon compounds differ considerably in the ease with which the molecules rotate around the carbon–carbon bonds. Rotation around the single bond in ethane requires very little energy, but rotation around the double bond in ethylene does not occur at an appreciable rate at temperatures below about 400°C. A rationalization of this restriction is the fact that rotation of the CH_2 groups with respect to each other would twist the atomic $2p(\pi)$ orbitals out of alignment, thus essentially breaking the π bond (see Figure 4–28).

Acetylene

In acetylene, C_2H_2, there is only one carbon atom and one hydrogen atom attached to either carbon atom. Thus we use two *sp* hybrid orbitals from each carbon atom to form σ bonds. For C_2H_2 there are three σ bonds, one C–C bond and two C–H bonds. In addition there are two $2p$ orbitals on each carbon atom that are available to form two π bonds. The resulting electronic structure of C_2H_2 is shown in Figure 4–30. The observed molecular structure of C_2H_2 is linear, which is consistent with the electronic structural description.

To summarize, the double bond in C_2H_4 consists of one σ and one π bond, and the triple bond in C_2H_2 consists of one σ and two π bonds. The relationship between bond order (the number of bonds between two atoms), bond distance, and bond energy is illustrated clearly in the series C_2H_6, C_2H_4, and C_2H_2. As

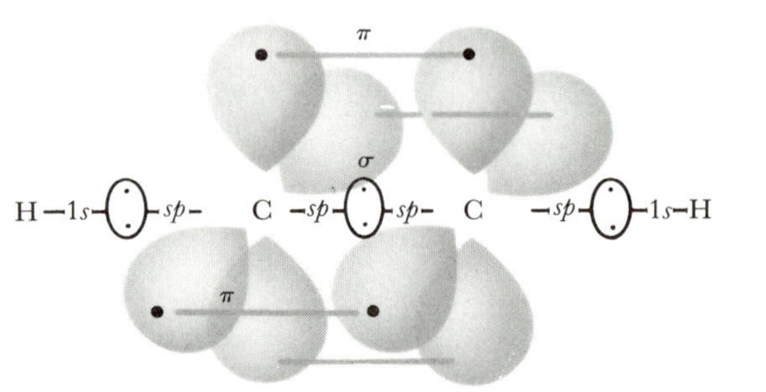

4–30
Localized-orbital representation of the bonding in C_2H_2 that shows the overlap of the two $2p$ orbitals on each carbon to form two π bonds.

the carbon-to-carbon bond order increases the bond length decreases, and the energy required to dissociate the $H_nC–CH_n$ bond into two H_nC fragments increases (see Table 4–3).

Table 4–3. Relationship of bond order to bond length and bond energy in three hydrocarbons

Molecule	C–C Bond order	C–C Bond length, Å	$H_nC–CH_n$ Bond energy, kcal $mole^{-1}$
C_2H_6	1	1.54	83
C_2H_4	2	1.35	125
C_2H_2	3	1.21	230

Benzene

The formula of benzene is C_6H_6, and the molecule has the planar structure shown in Figure 4–31. The planar hexagon of carbon atoms is an important molecular framework in structural chemistry, because it occurs in countless organic molecules. Therefore we will discuss the bonding and energy levels in the benzene molecule in detail.

Each carbon atom in the benzene ring is attached to two other carbon atoms and one hydrogen atom. Thus it is convenient to use sp^2 hybrid orbitals for each carbon atom to form the σ-bonding network shown in Figure 4–32. Since each carbon atom furnishes four valence electrons and each hydrogen atom furnishes one, there is one electron left for each carbon atom in a $2p$

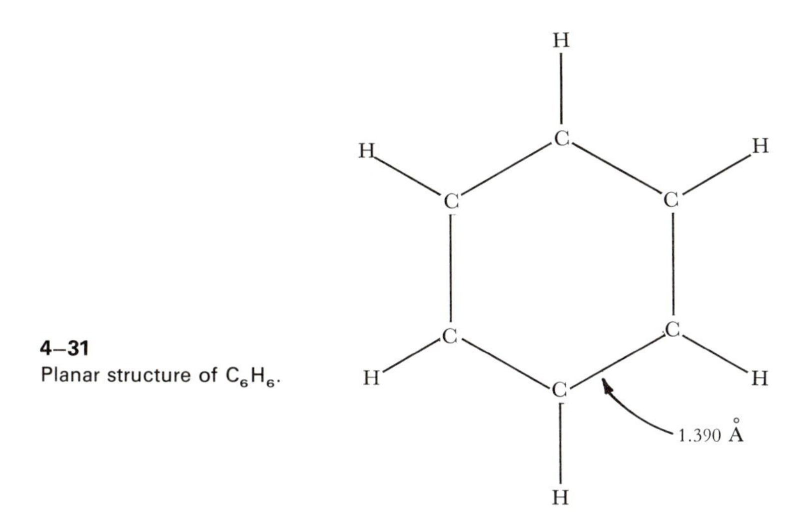

4–31
Planar structure of C_6H_6.

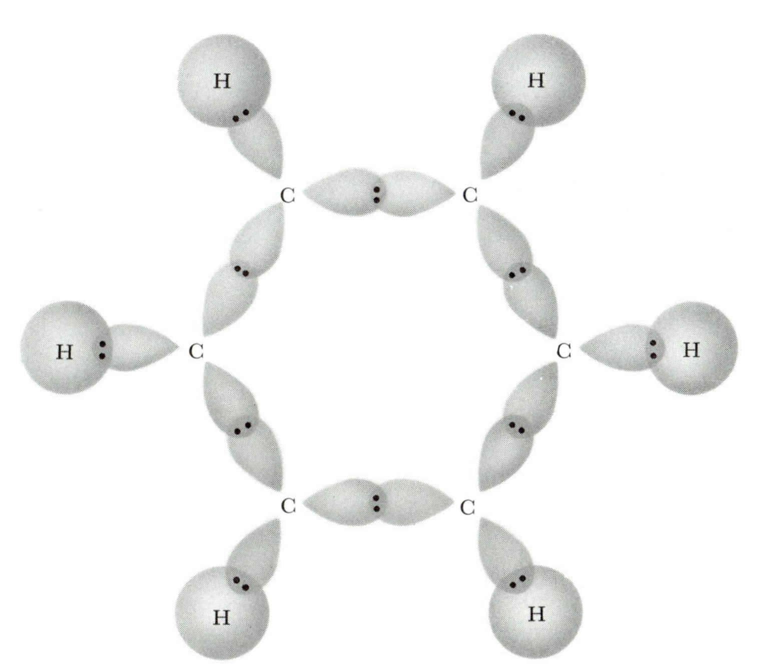

4–32
σ-Bond network in the benzene molecule.

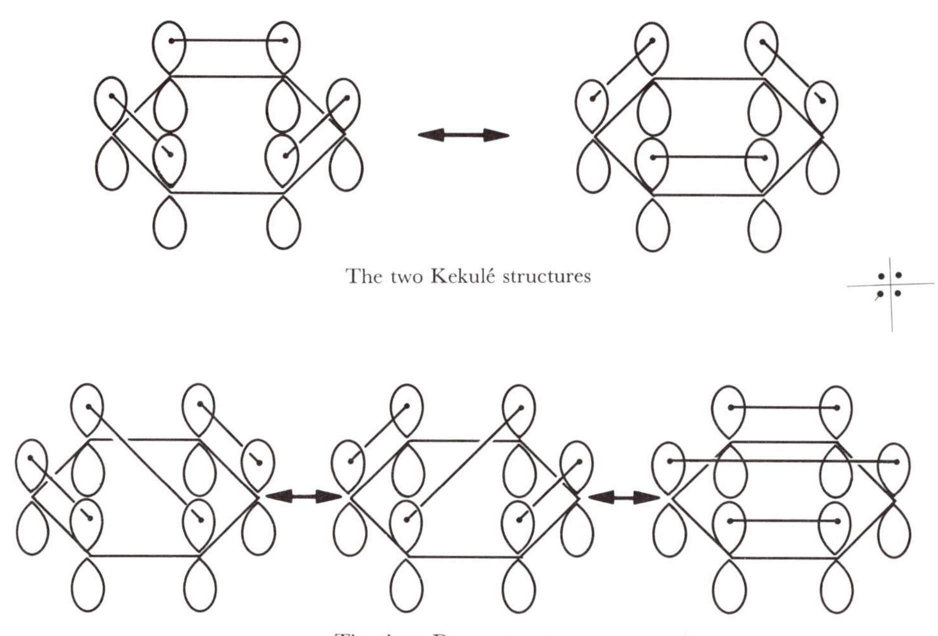

4–33
π-Bond resonance structures for the benzene molecule.

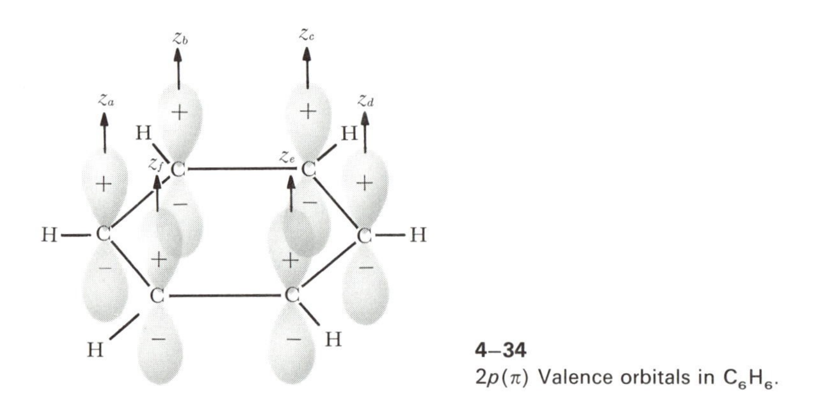

4–34
$2p(\pi)$ Valence orbitals in C_6H_6.

orbital perpendicular to the plane of the molecule. In a localized π-bonding scheme three localized electron-pair bonds can be formed in the ways shown in Figure 4–33, which depicts the localized π-bond structures.

The general expressions for the delocalized π molecular orbitals in benzene are given by appropriate linear combinations of the six $2p(\pi)$ valence orbitals.

Using the lettering system shown in Figure 4–34, we can formulate the bonding orbital of lowest energy as:

$$\psi(\pi_1^b) = \frac{1}{\sqrt{6}}(z_a + z_b + z_c + z_d + z_e + z_f)$$

The highest-energy antibonding orbital has nodes between the nuclei:

$$\psi(\pi_3^*) = \frac{1}{\sqrt{6}}(z_a - z_b + z_c - z_d + z_e - z_f)$$

The other molecular orbitals have energies between π_1^b and π_3^*:

$$\psi(\pi_2^b) = \frac{1}{2\sqrt{3}}(2z_a + z_b - z_c - 2z_d - z_e + z_f)$$

$$\psi(\pi_3^b) = \frac{1}{2}(z_b + z_c - z_e - z_f)$$

$$\psi(\pi_1^*) = \frac{1}{2\sqrt{3}}(2z_a - z_b - z_c + 2z_d - z_e - z_f)$$

$$\psi(\pi_2^*) = \frac{1}{2}(z_b - z_c + z_e - z_f)$$

The π molecular orbitals for benzene are shown in Figure 4–35. A calculation of the energies of the six π molecular orbitals gives the energy-level scheme for C_6H_6 that is shown in Figure 4–36.

There are a total of 30 valence electrons in benzene. Twenty-four electrons are used in σ bonding (six C–C bonds and six C–H bonds), thereby leaving six electrons for the π molecular orbitals shown in Figure 4–36. In the ground state the π electrons have the configuration $(\pi_1^b)^2(\pi_2^b)^2(\pi_3^b)^2$, thereby giving a total of three π bonds. Therefore each carbon–carbon bond consists of one full σ bond and half a π bond. The carbon–carbon bond length in C_6H_6 is 1.390 Å, which is between the C—C and C═C bond lengths.

Simplified representations for the bonding in benzene are shown in Figure 4–37. Both the localized and delocalized bonding pictures are given for comparison. Benzene actually is more stable than might be expected for a system of six C–C single bonds and three C–C π bonds. This added stability is due to the delocalization of the electrons in the three π bonds over all six carbon atoms, as is evident from the molecular orbitals shown in Figure 4–35. If we did not allow the delocalization of electrons in C_6H_6, we would have a system of three isolated double bonds (only one of the Kekulé structures shown in Figure 4–37). An electron in a localized π^b orbital of a C═C bond has less energy than an electron in a carbon $2p$ atomic orbital by an amount equal to that of an electron in a π_2^b or π_3^b molecular orbital of benzene. Because two electrons are placed in the π_1^b orbital the delocalization of three π bonds in C_6H_6 gives an added stability equal to twice the energy difference of the π_2^b and

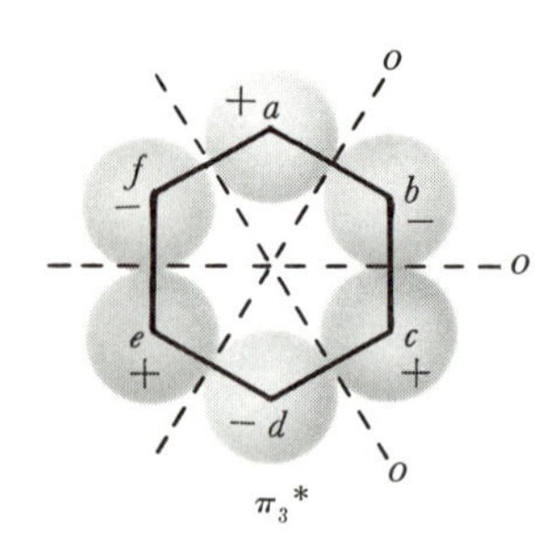

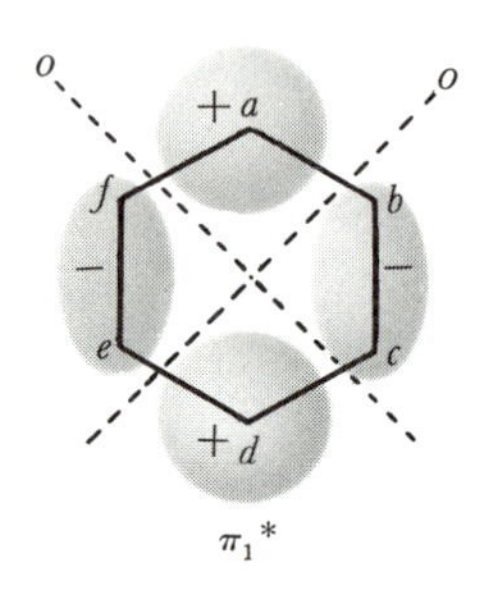

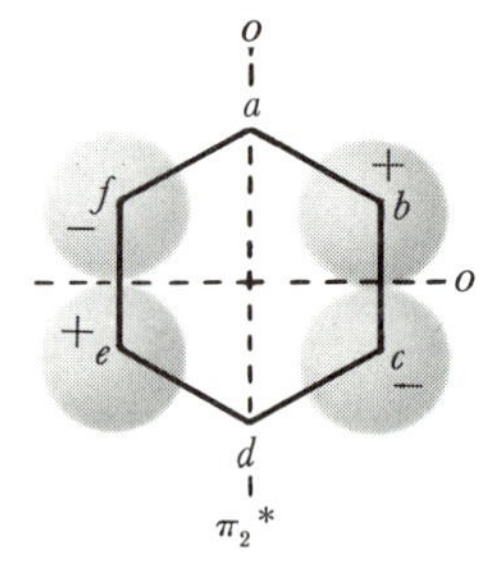

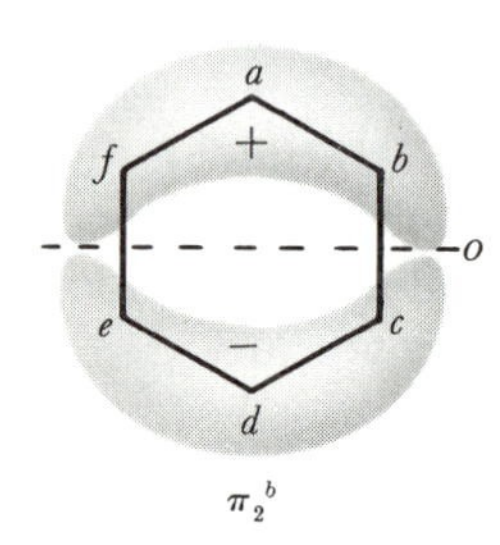

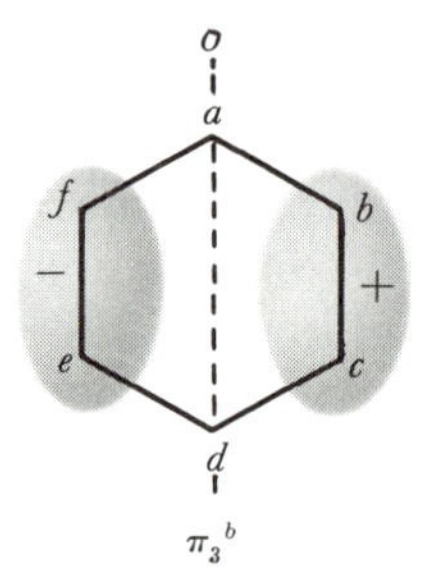

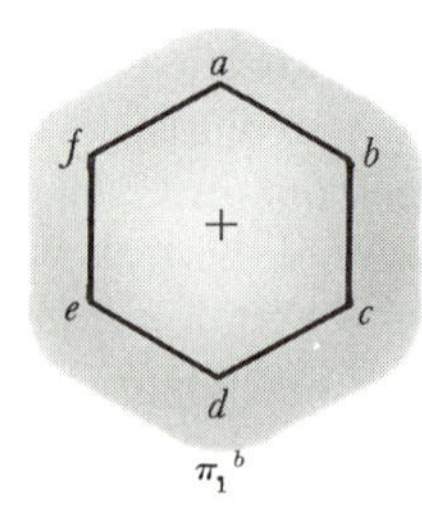

4–35
π Molecular orbitals for benzene.

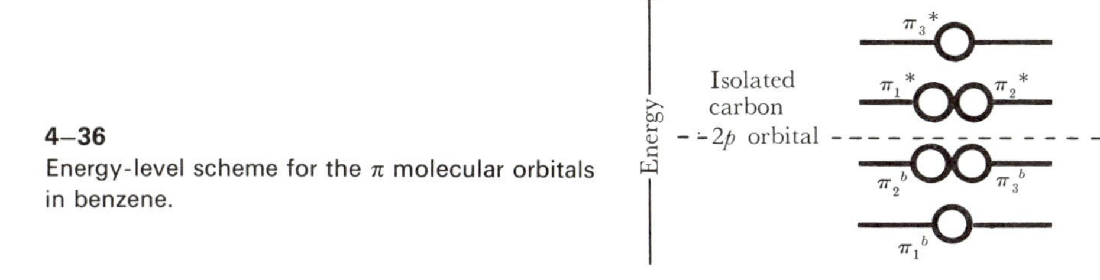

4–36
Energy-level scheme for the π molecular orbitals in benzene.

Kekulé structures

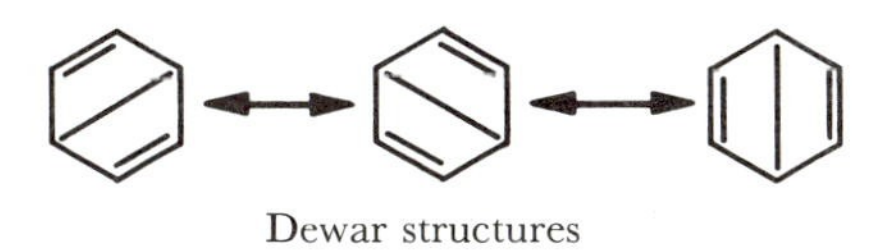

Dewar structures

Simple molecular-orbital picture

4–37
Simplified representations of bonding in a benzene molecule.

π_1^b levels. This "*resonance*" *energy* is computed to be equal to the energy of one isolated C–C π bond.

The so-called experimental resonance energy of benzene is obtained by adding the bond energies of the C–C, C═C, and C–H bonds and comparing the total with the experimentally known value for the heat of formation of benzene. The difference indicates that benzene is about 40 kcal mole^{-1} lower in energy than the sum of the bond energies for a system of six C–H, three C–C, and three isolated C═C bonds would suggest.

4–8 MOLECULAR SPECTROSCOPY

In addition to electronic energy levels, molecules possess energy levels associated with rotational (Figure 4–38) and vibrational motion (Figure 4–39). Generally, any linear polyatomic molecule can rotate around the three mutually perpendicular axes through the center of gravity of the molecule, as shown in Figure 4–38. For the case of a linear molecule (a diatomic molecule must

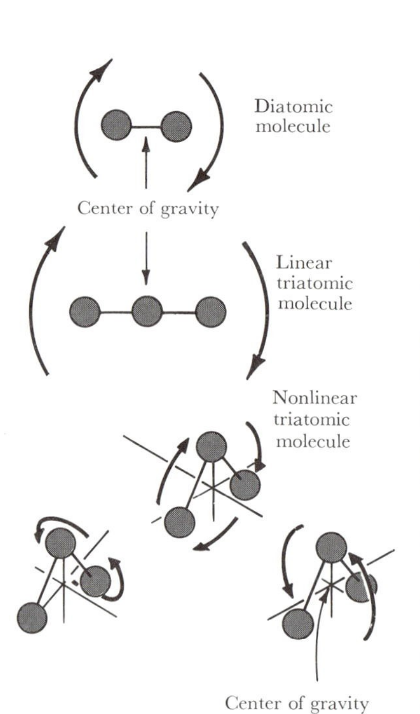

4–38
Rotational motion in molecules.

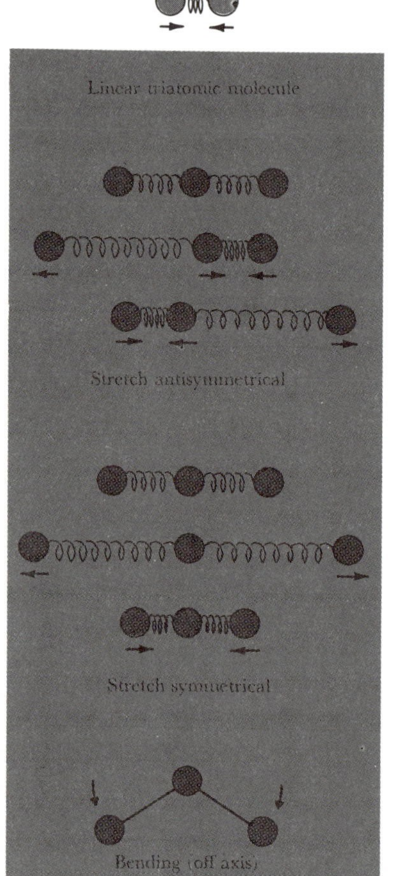

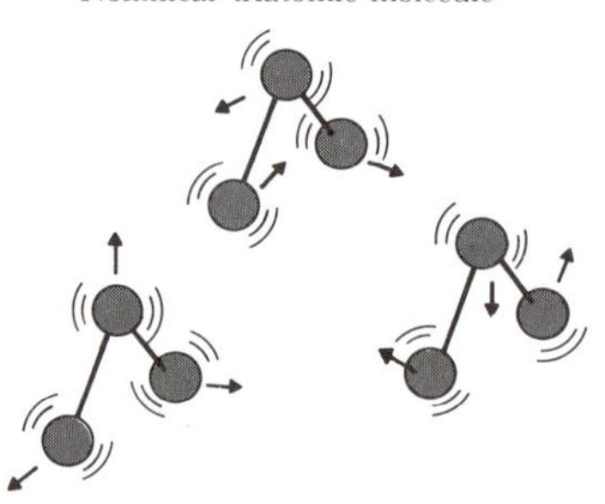

4–39
Vibrational motion in molecules.

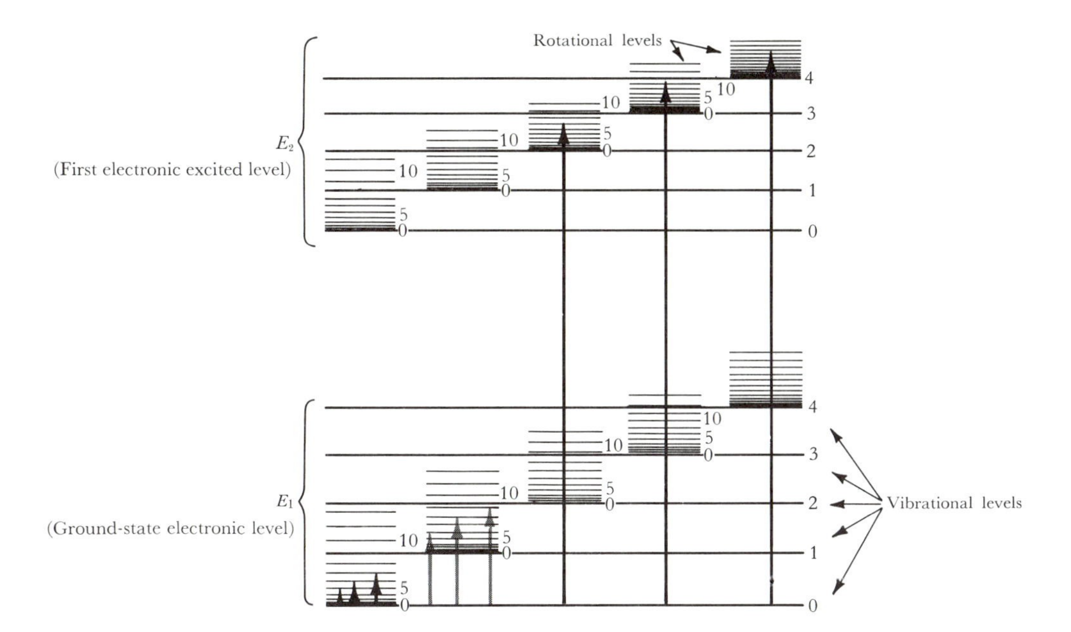

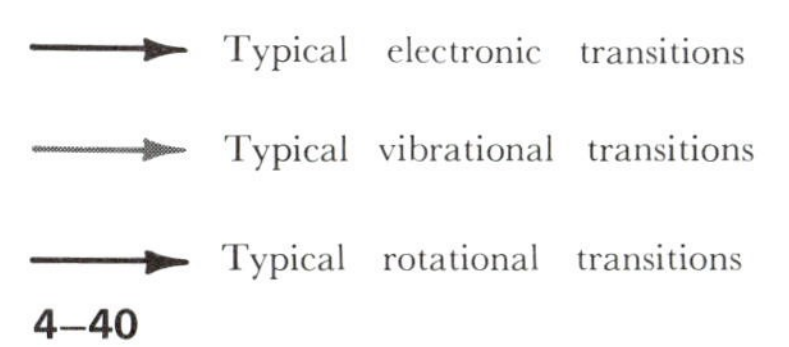

4–40
Generalized energy-level diagram for a molecule.

be linear), one of these axes lies on the line between the centers of the atoms, thus there are only two ways of rotation. Figure 4–39 shows the modes of vibration of diatomic, linear triatomic, and nonlinear triatomic molecules. Often it is helpful in discussing vibrations in molecules to treat the bonds between the atoms as springs, as shown.

A reasonable approximation for the total molecular energy is

$$E_{\text{total}} = E_{\text{electronic}} + E_{\text{vibrational}} + E_{\text{rotational}}$$

Figure 4–40 shows a generalized energy-level diagram for a molecule. Two electronic levels, E_1 and E_2, are shown with their vibrational and rotational levels. Separations between electronic energy levels usually are much larger than those between vibrational levels, which in turn are much larger than those between rotational levels. *Electronic transitions* correspond to absorption of radiation in the visible and ultraviolet portion of the spectrum, *vibrational*

Table 4–4. Electromagnetic radiation and spectroscopy

Frequency, sec^{-1}	10^{6}	10^{7}	10^{8}	10^{9}	10^{10}	10^{11}
Wave number, cm^{-1}	3.3×10^{-5}	3.3×10^{-4}	0.0033	0.0333	0.333	3.33
Wavelength	300 m	30 m	3 m	30 cm	3 cm	0.3 cm
Energy	4000 erg $mole^{-1}$	0.004 joule $mole^{-1}$	0.04 joule $mole^{-1}$ 0.01 cal $mole^{-1}$	0.4 joule $mole^{-1}$ 0.1 cal $mole^{-1}$	4 joule $mole^{-1}$ 1 cal $mole^{-1}$	40 joule $mole^{-1}$ 10 cal $mole^{-1}$
Name	Radio ⟶ (long-wave)	(short-wave)	(television and FM) (UHF)	⟶ Microwave ⟶		⟶ Infrared ⟶ (far IR)
Source and detector	(not used)	Vacuum tubes, wires, antenna, coil		Klystron, guide, cavity		Hot wire, etc.
Frequency, sec^{-1}	10^{12}	10^{13}	10^{14}	10^{15}	10^{16}	10^{17}
Wave number, cm^{-1}	33	333	3333	33333	3.3×10^{5}	3.3×10^{6}
Wavelength	300 μ	30 μ	3 μ 30,000 Å	300 nm 3000 Å	30 nm 300 Å	3 nm 30 Å
Energy	0.1 kcal $mole^{-1}$ 0.16 kT	1 kcal $mole^{-1}$ 1.6 kT	10 kcal $mole^{-1}$ 16 kT	100 kcal $mole^{-1}$ 4 eV	1000 kcal $mole^{-1}$ 40 eV	400 eV
Name	Infrared ⟶	(near IR)	⟶ Visible ⟶ (red ⟶ blue)	⟶ Ultraviolet ⟶ (near UV)		⟶ (far UV)
Source and detector	Lamp, prism, grating; phototube or photographic plate			Lamp, etc., grating, phototube		
Frequency, sec^{-1}	10^{18}	10^{19}	10^{20}			
Wave number, cm^{-1}	3.3×10^{7}	3.3×10^{8}	3.3×10^{9}			
Wavelength	3 Å	0.3 Å	0.03 Å			
Energy	4 keV	40 keV	400 keV 0.4 MeV			
Name	X rays ⟶			⟶ Gamma rays		
Source and detector	X-ray tube, photographic plate			Nuclear reaction, counter		

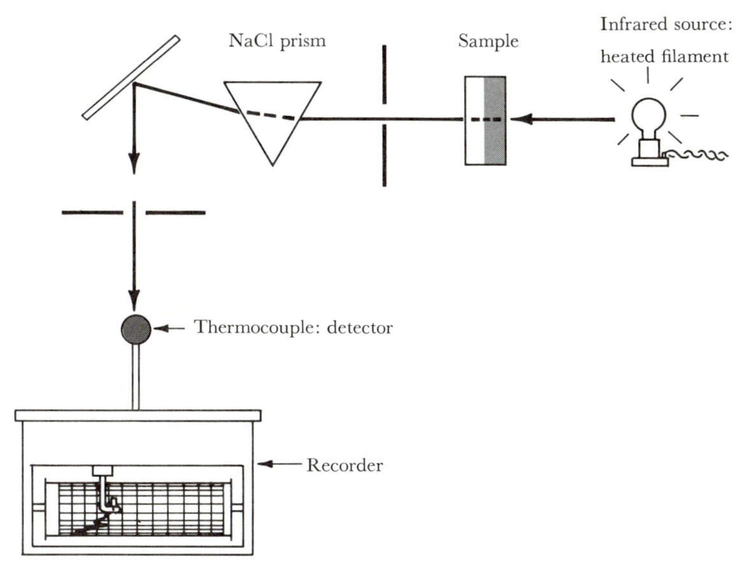

4–41
Schematic of an infrared spectrophotometer.

transitions correspond to absorption in the near-infrared and infrared regions, and *rotational transitions* correspond to absorption in the far-infrared to microwave regions.

Table 4–4 shows the range of electromagnetic radiation, gives energies in all commonly used units, and indicates the kind of spectroscopy that is used in each range. The quantized properties of molecular energy levels have been utilized in modern spectroscopic techniques to help to identify molecules and to assign molecular structures. For example, the study of rotational transitions by far-infrared and microwave spectroscopy is used to obtain extremely precise data on bond angles and bond lengths.

In infrared spectroscopy a beam of infrared radiation, whose wavelength varies from 2.5 μ to 15 μ (wave number varies from 4000 cm^{-1} to 667 cm^{-1}) is passed through a sample of a compound (Figure 4–41). Often, the sample first is compressed under pressure into a thin wafer and inserted into infrared-transparent sodium chloride sample holders. Sodium chloride must be used because quartz or glass is opaque to infrared light. The sodium chloride prism position and the slit widths determine the wavelength of the radiation reaching the detector. The absorption of radiation at different wave numbers corresponds to the excitation of molecules from a low (usually the lowest) vibrational energy level to the next higher vibrational energy level.

Absorbed radiation is identified by its wavelength (λ in Å, μ, or nm; $1\ \mu = 10^3$ nm $= 10^4$ Å $= 10^{-4}$ cm), its frequency (ν in sec^{-1}), or its wave number ($\bar{\nu}$ in cm^{-1}). Radiation absorption is detected electronically and recorded in some suitable form as a graphical trace. A strong absorption throughout a narrow range of frequencies causes a sharp "peak" or "line" in the recorded spectrum. Absorption peaks are not always narrow and sharp because each vibrational energy level has superimposed upon it an array of rotational energy levels (Figure 4–40); therefore a particular vibrational transition is really the superposition of transitions from many vibrational-rotational levels.

Different absorbed radiation frequencies correspond to different intramolecular excitations. For example, stretching vibrations for C=O bonds occur in the region below 6 μ (1780 cm^{-1}–1850 cm^{-1}) and for C–H bonds in the region between 3 μ and 4 μ (2800 cm^{-1}–3000 cm^{-1}). Absorption peaks in other regions of the infrared (IR) spectrum correspond to energy changes in other bonds or to complex intramolecular vibrations. Table 4–5 lists the characteristic IR absorption wave numbers of certain bonds.

Table 4–5. Positions of some characteristic infrared bands of molecular groups

Bond	Group	Description	$\bar{\nu}$, cm^{-1}	λ, μ
C—H	CH_2, CH_3	Stretching	3000–2900	3.3–3.4
C—H	≡C—H	Stretching	3300	3.0
C—H	Aromatic (benzene)	Stretching	3030	3.3
C—H	—CH_2—	Bending	1465	6.8
C—H	—CH=CH— (trans)	In-plane bending	970–960	10.3–10.4
		Stretching	1310–1295	7.6–7.7
C—H	—CH=CH— (cis)	Out-of-plane bending	~690	14.5
O—C	>C=O	Stretching	1850–1700	5.4–5.9
O—H	—O—H[a] (alcohols)	Stretching	3650–3590	2.7–2.8
O—H	Hydrogen bonded (alcohols)	Stretching?	3400–3200	3.0–3.1

[a] Refers to an O–H group that is not hydrogen bonded to a neighboring molecule. Hydrogen bonding is discussed in Section 6–2.

Infrared spectroscopy is extremely useful in the identification of unknown materials because each chemical compound has a unique IR spectrum. Figure 4–42 shows the IR spectra of tetrachloroethylene and cyclohexene.

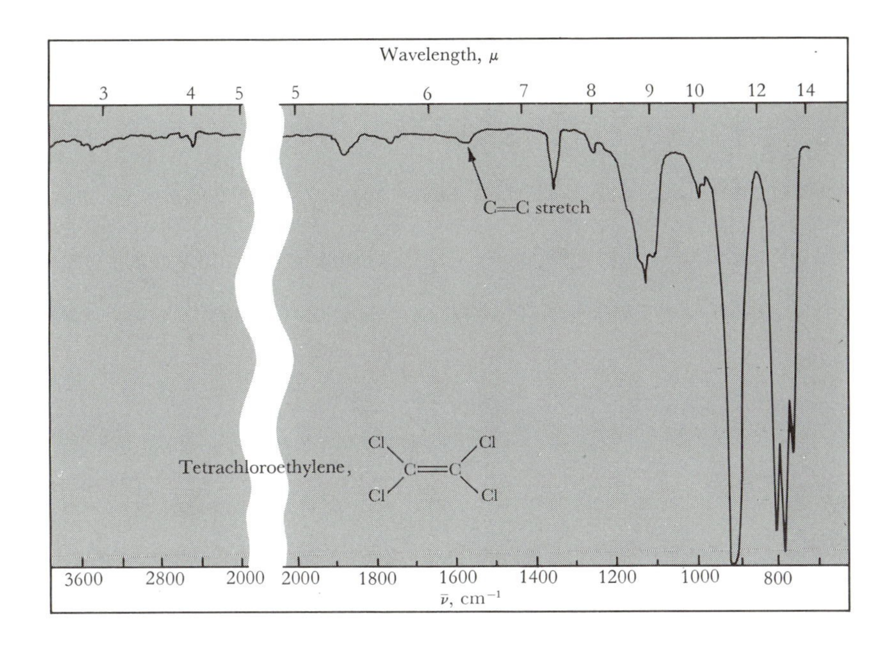

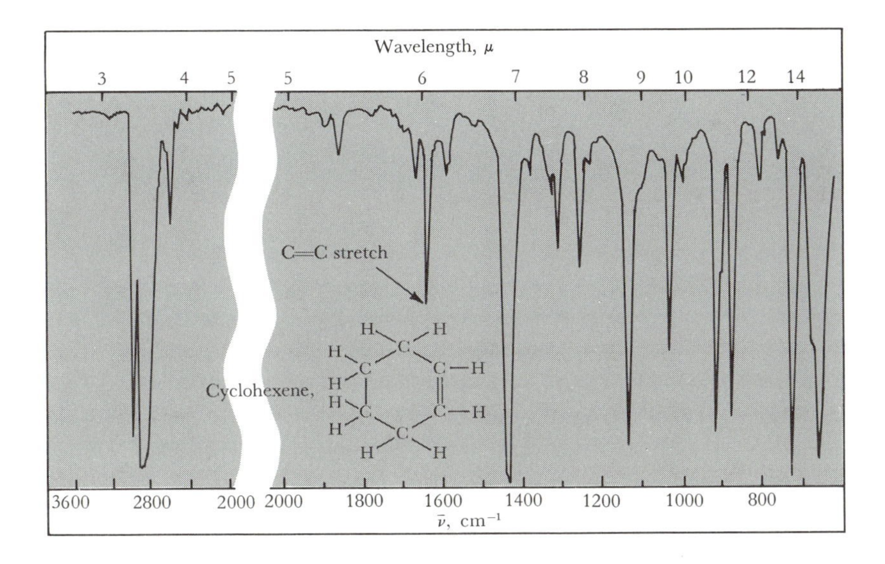

4-42
Infrared spectra of tetrachloroethylene and cyclohexene.

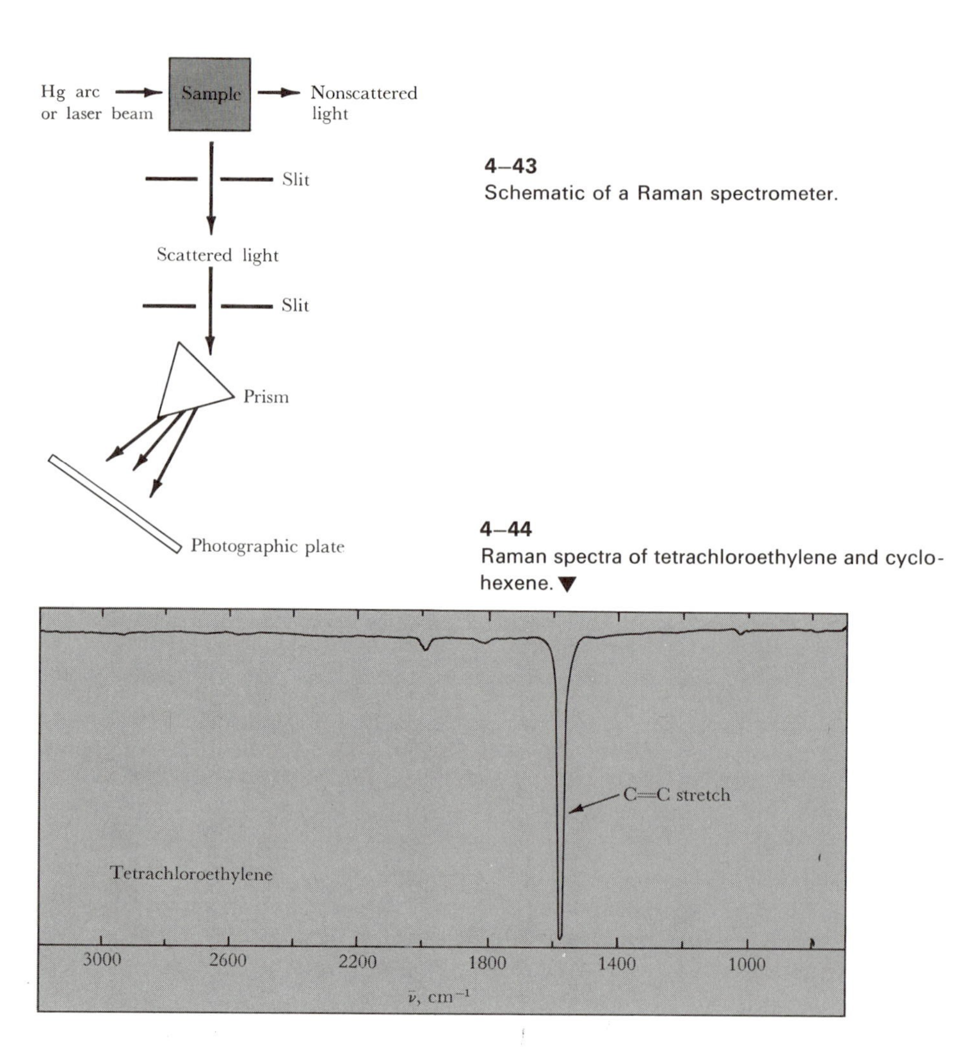

4–43
Schematic of a Raman spectrometer.

4–44
Raman spectra of tetrachloroethylene and cyclohexene. ▼

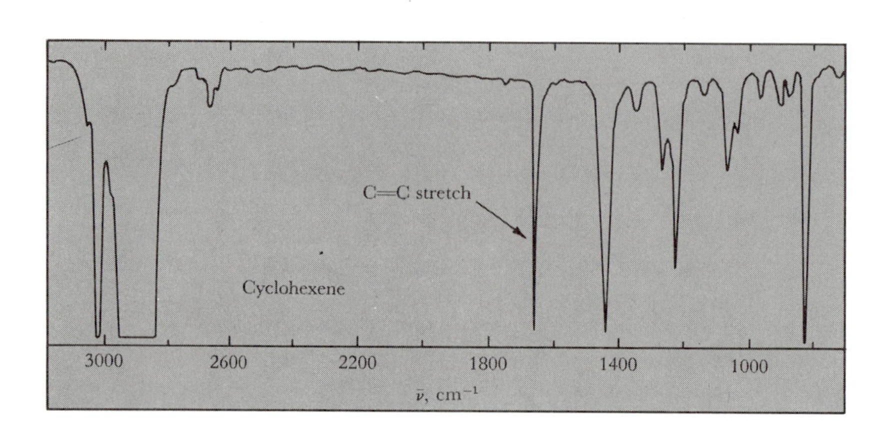

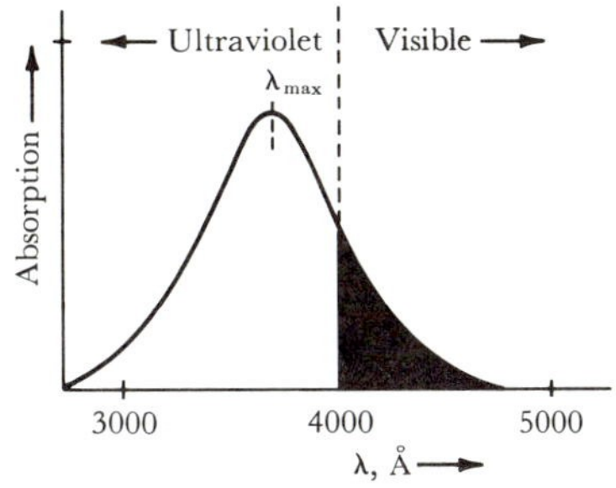

4–45
A typical electronic absorption band with λ_{max} in the near-ultraviolet region.

Not all molecules respond to infrared radiation. In particular, molecules with certain elements of symmetry, such as homonuclear diatomic molecules, do not absorb infrared radiation. In more complex molecules not all modes of vibration respond to infrared radiation. For example, symmetrical molecules such as ethylene, $H_2C{=}CH_2$, do not exhibit all their vibrations in infrared spectra. To assist in examining the vibrations of these molecules, *Raman spectroscopy* often can be used. The Raman spectrum results from the irradiation of molecules with light (usually in the visible region) of a known wavelength. A laser beam commonly is used to irradiate the sample in Raman spectrometers (Figure 4–43). Radiation absorption is not detected directly. When irradiated with light of high energy, the molecules may add to, or extract from, the incident light small amounts of energy corresponding to the energy of some particular molecular vibration. The incident light is said to be scattered, rather than absorbed, and may be observed as light with a wavelength different from the incident light. Figure 4–44 shows the Raman spectra of tetrachloroethylene and cyclohexene.

As stated previously, electronic transitions correspond to the absorption of large amounts of energy, compared to the energy absorbed in vibrational or rotational transitions. Electronic excitations usually are associated with absorption of visible and ultraviolet light. Just as vibrational absorption "bands" are widened by a superposition of many vibrational-rotational transitions, absorption spectra recorded in the visible–ultraviolet region exhibit broad bands, rather than sharp peaks, because of the superposition of many vibrational-electronic transitions (Figure 4–45). Absorption bands are characterized by a maximum at a particular wavelength, λ_{max}.

A schematic diagram of a visible and ultraviolet spectrophotometer is shown in Figure 4–46. A prism, or diffraction grating, and the width of the slit regulate the spread of wavelengths through the sample. The sample in solution is compared to a reference blank containing pure solvent. The light incident on the sample is denoted I_0 and the transmitted light is denoted I. The absorption of light can be expressed quantitatively by the relationship known as *Beer's law*, $A = \varepsilon c l$, in which A is the absorbance ($\log I_0/I$), c is the

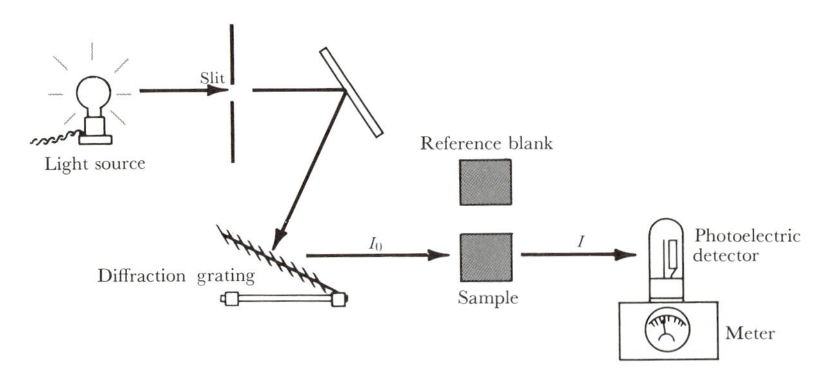

4–46
Schematic diagram of a visible and ultraviolet spectrophotometer.

concentration of the substance (in moles liter^{-1}), and l is the length (in cm) of the light path through the substance. The *molar extinction coefficient*, ε, is characteristic of the absorbing sample. Values for the molar extinction coefficient vary considerably from compound to compound and from peak to peak in the absorption spectrum of the same compound. The extent to which an electronic transition is "allowed" or "forbidden" by quantum mechanical selection rules is reflected in the value of ε of the absorption peak accompanying that transition. Thus the experimentally determined ε value often can be extremely useful in assigning a given absorption peak to a particular type of electronic transition. If ε is in the range of 10^3–10^5, the transition meets all the requirements of the quantum mechanical selection rules. The excitation of an electron in the π bonding orbital to the π antibonding orbital of C_2H_4, which was discussed in Section 4–7, is an example of such a transition. The ε value of the $\pi^b \rightarrow \pi^*$ band in C_2H_4 is approximately 10^4. An important example of a forbidden transition is the excitation of a nonbonding $2p$ oxygen electron in molecules containing the carbonyl group (C═O) to the π^* orbital. This commonly is called a $n \rightarrow \pi^*$ transition, which is said to be orbitally forbidden.

The selection rules are only approximate when they describe a transition as forbidden because in actual practice an absorption band generally can be observed. Even so, the intensity of the band will be diminished significantly if it corresponds to an orbitally forbidden transition; in this situation ε usually is in the range 10^0–10^3.

Finally, extremely small values of ε (10^{-5}–10^0) are associated with an electronic transition in which the spin of the electron is required to change in going from the ground state to the excited state. Thus a "spin-forbidden" transition often is difficult to observe as an absorption band because of its extraordinarily weak intensity.

SUGGESTIONS FOR FURTHER READING

C. J. Ballhausen and H. B. Gray, *Molecular Orbital Theory*, Benjamin, Menlo Park, Calif., 1964.

E. Cartmell and G. W. A. Fowles, *Valency and Molecular Structure*, 3rd ed., Butterworths, London, 1966.

A. Companion, *Chemical Bonding*, McGraw-Hill, New York, 1964.

C. A. Coulson, *Valence*, 2nd ed., Oxford, New York, 1961.

H. B. Gray, *Electrons and Chemical Bonding*, Benjamin, Menlo Park, Calif., 1965.

M. Karplus and R. N. Porter, *Atoms and Molecules: An Introduction for Students of Physical Chemistry*, Benjamin, Menlo Park, Calif., 1970.

L. Pauling, *The Chemical Bond*, Cornell Univ. Press, Ithaca, N.Y., 1967.

G. C. Pimentel and R. D. Spratley, *Chemical Bonding Clarified Through Quantum Mechanics*, Holden-Day, San Francisco, 1969.

QUESTIONS AND PROBLEMS

1. Indicate an appropriate hybridization for the central-atom valence orbitals and predict the molecular shape and polarity for each of the following molecules:

   a) CS_2 b) SO_3 c) ICl_3 d) BF_3
   e) CBr_4 f) SiH_4 g) SF_2 h) SeF_6
   i) PF_3 j) ClO_2 k) IF_5 l) OF_2
   m) H_2Te

2. For the molecules and ions CO_2, NO_2^+, NO_2, NO_2^-, and SO_2: (a) Represent their electronic structures by drawing localized molecular-orbital structures. (b) Predict their molecular shapes and indicate which neutral molecules are polar. (c) Predict the number of unpaired electrons for each molecule. (d) Predict the bond angles in NO_2^+ and NO_2^-. (e) Predict relative N–O bond lengths for NO_2^+, NO_2, and NO_2^-.
3. Draw a localized molecular-orbital structure for N_3. What is its molecular shape? Discuss the electronic structure of N_3. How many unpaired electrons are there? Is the molecule polar? Predict the relative N–N bond lengths in N_2 and N_3.
4. Use VSEPR theory to predict the molecular shapes of XeF_4, XeO_4, XeO_3, and XeF_2.
5. Give the orbital hybridization of the central atom and the VSEPR-theory geometrical structure for AsH_3, ClF_3, and $SeCN^-$. In the case of $SeCN^-$ draw a localized molecular-orbital representation of its electronic structure, showing the σ bonds, π bonds, and lone electron pairs.
6. Use VSEPR theory to formulate the hybridization of the central atom (e.g., *sp*, sp^2, sp^3d) and the molecular geometry (e.g., bent, linear, pyramidal) of each of the following molecules and ions: (a) SO_3^{2-}; (b) SO_3; (c) BrF_5; (d) I_3^-; (e) CH_3^+; (f) CH_3^-; (g) PCl_3F_2.
7. Which of the neutral molecules in Problem 6 would you expect to have dipole moments?

8. Borazine has the formula $(BHNH)_3$. The combination of a boron atom and a nitrogen atom is isoelectronic with two carbon atoms and has the same sum of atomic weights. How would you formulate the electronic structure of borazine?
9. A compound is found to have the empirical formula $(NH_4)_2SbCl_6$. X-ray diffraction studies reveal that the anion consists of discrete units containing one Sb and six Cl's. The compound is diamagnetic. The electronic configuration of Sb is $(Kr)5s^25p^3$. Would you expect the compound to be only $(NH_4)_2SbCl_6$, or an equal mixture of $(NH_4)_3SbCl_6$ and $(NH_4)SbCl_6$? Why?
10. Attempt an explanation of the following dipole moments in terms of a localized-orbital model of electronic structure: NH_3, $\mu = 1.47$ D; PH_3, $\mu = 0.55$ D; NF_3, $\mu = 0.23$ D.
11. Three geometrical structures commonly are found for three-coordinate molecules of the main-group elements. What are these structures? In each case how many lone electron pairs does the central atom have, according to VSEPR theory? Using chlorine atoms bonded to a central atom give an example of each structure. Which molecules among these examples are polar?
12. Formulate a localized-orbital model for both the σ and π bonds in the ozone molecule, O_3. What is the hybridization of the central oxygen atom? How many π bonds are there between each pair of oxygen atoms?
13. Tellurium (Te) is a rather vile element, so only a few chemists work with it. To inspire the handful of tellurium chemists to greater achievement, predict the formulas of the tellurium-fluorine molecules (or ions) which would exhibit the following geometrical structures: (a) angular; (b) T-shaped; (c) trigonal pyramidal; (d) sawhorse; (e) square planar; (f) square pyramidal; (g) trigonal bipyramidal; (h) octahedral.

    Predict formulas of tellurium-oxygen molecules and ions that are: (a) angular; (b) trigonal pyramidal; (c) trigonal planar; (d) tetrahedral.
14. Predict the shapes of the following ions, using VSEPR theory: (a) AlF_6^{3-}; (b) TlI_4^{3-}; (c) $GaBr_4^-$; (d) NO_3^-; (e) NCO^-; (f) CNO^-; (g) $SnCl_3^-$; (h) $SnCl_6^{2-}$.
15. Draw localized-orbital representations including π bonds for NO_3^-, NCO^-, and CNO^-.
16. The allyl cation has the formula $C_3H_5^+$. What structure does it have? Formulate a localized-orbital model of the electronic structure of $C_3H_5^+$. Repeat the problem for the allyl radical, C_3H_5, and the allyl anion, $C_3H_5^-$.
17. Formaldehyde (H_2CO) exhibits a strong electronic absorption band in the ultraviolet region, which may be assigned to a $\pi^b \rightarrow \pi^*$ transition (see the discussion of ethylene). In addition, a weaker, longer-wavelength peak is observed in the spectra of H_2CO, and all organic compounds containing a carbonyl (C=O) group, in the region 2700–3000 Å. Formulate the molecular orbitals for H_2CO and suggest a possible assignment for the long-wavelength peak.
18. Which molecule would you expect to have the higher ionization energy, ethylene or ethane? Why? Also predict the relative *IE*'s of ethylene and acetylene.
19. Is XeF_5^+ a reasonable species? What geometry would you predict it to have?
20. Would you expect benzene to absorb light of lower, or higher, energy than ethylene? Why?

5

Transition-Metal Complexes

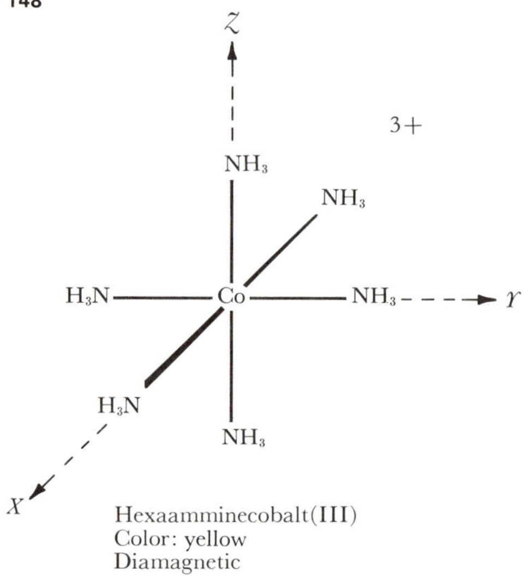

Hexaamminecobalt(III)
Color: yellow
Diamagnetic

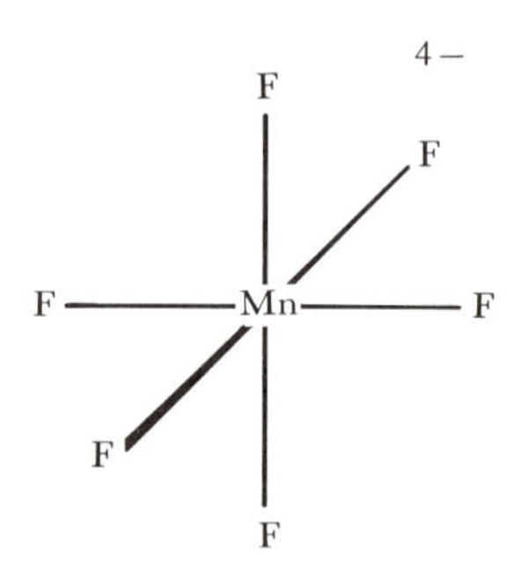

Hexafluoromanganate(II)
Color: pink
Paramagnetic (5 unpaired electrons)

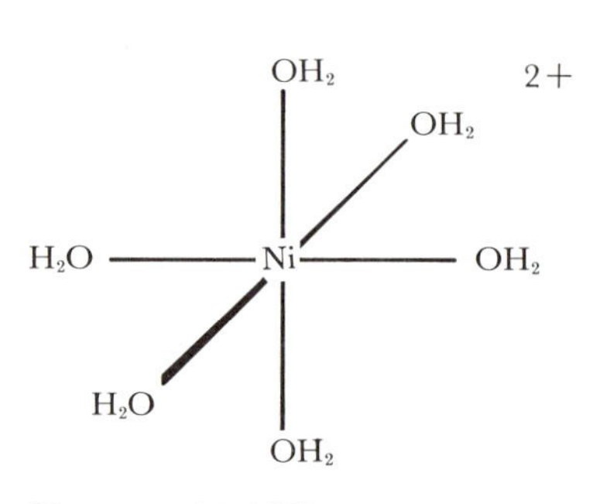

Hexaaquonickel(II)
Color: green
Paramagnetic (2 unpaired electrons)

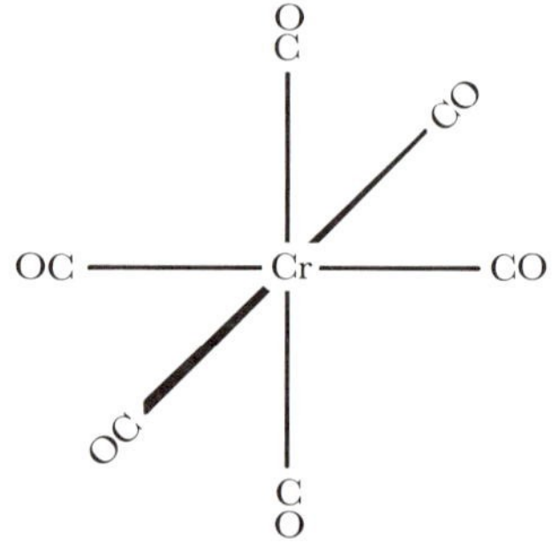

Hexacàrbonylchromium(0)
Color: white
Diamagnetic

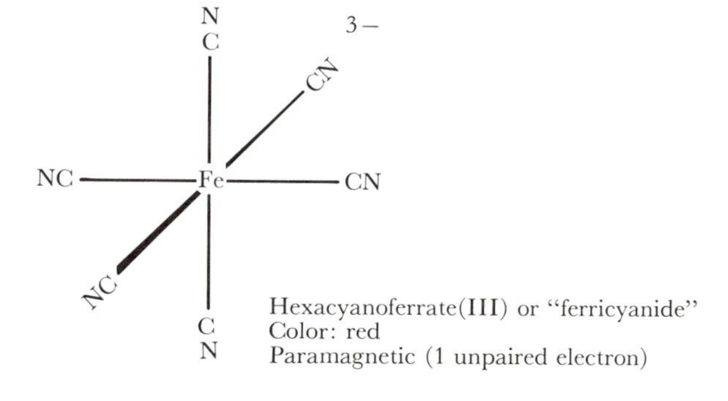

Hexacyanoferrate(III) or "ferricyanide"
Color: red
Paramagnetic (1 unpaired electron)

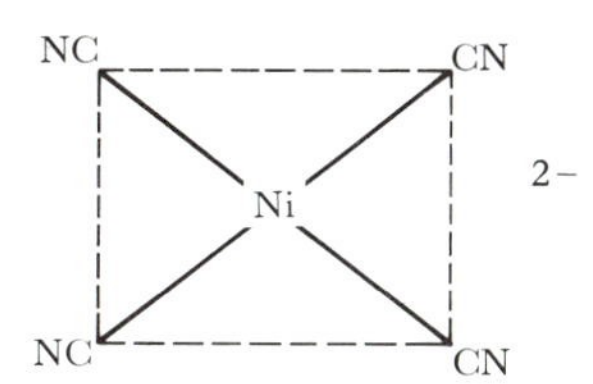

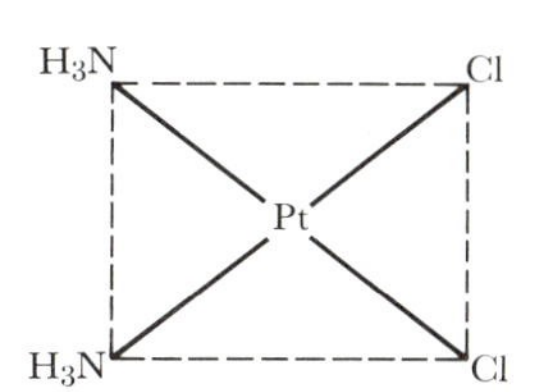

Square planar

Tetracyanonickelate(II)
Color: yellow
Diamagnetic

cis-Dichlorodiammineplatinum(II)
Color: yellow
Diamagnetic

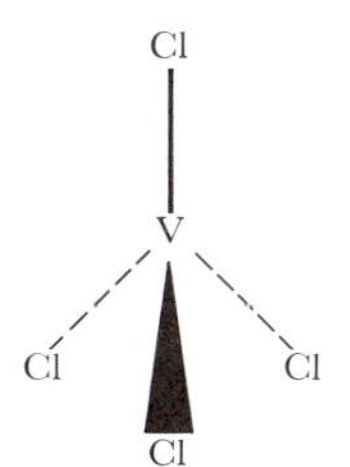

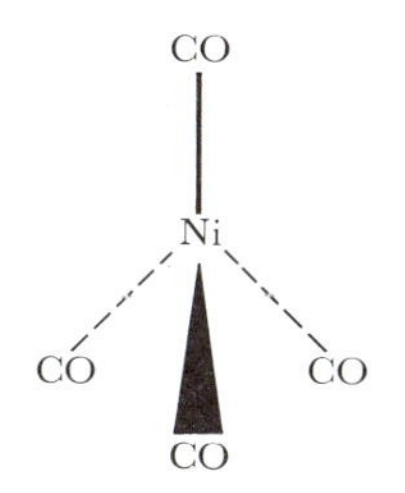

Tetrahedral

Tetrachlorovanadium(IV)
Color: red-brown
Paramagnetic (1 unpaired electron)

Tetracarbonylnickel(0)
Color: colorless
Diamagnetic

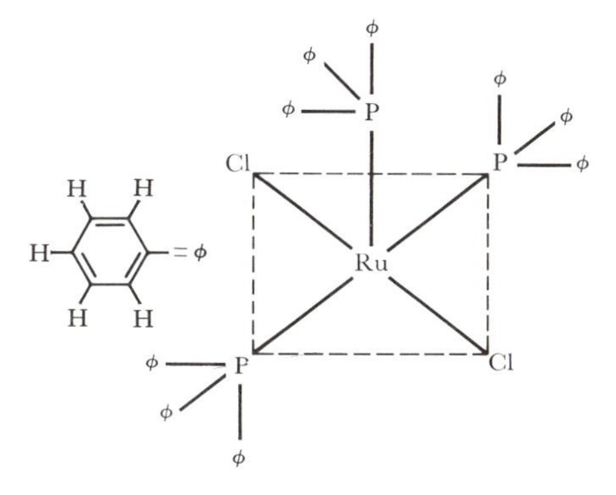

Square pyramidal

trans-Dichloro(tris)triphenyl-
phosphineruthenium(II)
Color: brown
Diamagnetic

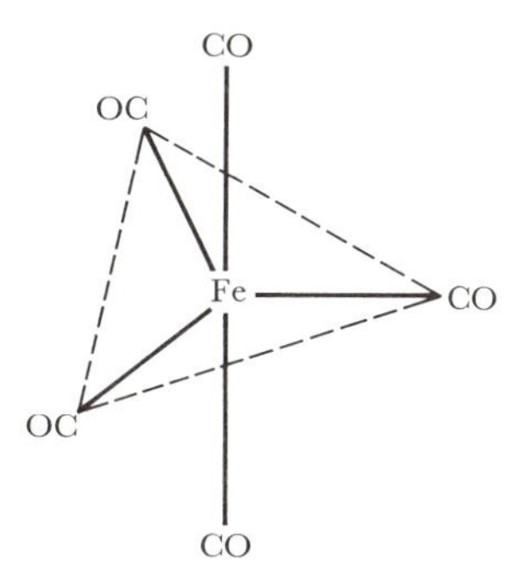

Trigonal bipyramidal

Pentacarbonyliron(0)
Color: yellow
Diamagnetic

5–1
Several metal complexes, with systematic names, colors, and magnetic properties.

The transition metals (or *d*-electron elements) form a wide variety of compounds with interesting spectroscopic and magnetic properties. Many of these compounds, or complexes as they often are called, are catalysts in important industrial chemical processes. Furthermore, several transition-metal ions are necessary components of biochemical systems. The electronic structures of transition-metal complexes have been studied extensively in the past twenty years.

In discussing the structures of metal complexes it is convenient to use the term *coordination number*, which is defined as the number of atoms bonded directly to the central metal atom. The groups attached to the metal are called *ligands*. Each ligand has one or more donor atoms that bond to the metal. Emphasis has been on coordination numbers four and six, which are by far the most common among transition-metal complexes. Almost all of the six-coordinate complexes have an octahedral structure. Both square planar and tetrahedral geometries are prominent for four-coordinate complexes. A large number of complexes of metal ions with d^8 or d^9 electronic configurations have the square planar structure. For example, most of the Pd^{2+} ($4d^8$), Pt^{2+} ($5d^8$), and Au^{3+} ($5d^8$) complexes are square planar. Although tetrahedral complexes are formed by many metal ions of the first transition series (Sc–Zn), the occurrence of this structure in heavier (4*d* and 5*d* series) metal ions is restricted mainly to d^0 and d^{10} configurations (e.g., MoO_4^{2-}, ReO_4^-, and HgI_4^{2-}).

Figure 5–1 shows examples of the most important geometries for coordination numbers four, five, and six. Recent research has shown the importance of the once rare square pyramidal and trigonal bipyramidal five-coordinate structures in the ground-state stereochemistry of many metal ions, particularly that of Ni^{2+}. Other lesser known coordination geometries, based on coordination numbers greater than six, now are receiving much attention.

5–1 *d* ORBITALS IN BONDING

The maximum number of σ bonds that can be constructed from *s* and *p* valence orbitals is four. Thus four is the highest coordination number commonly encountered for central atoms with 2*s* and 2*p* valence orbitals. For example, in CH_4 the central carbon atom is "saturated" with four σ bonds. However, with a first-row transition metal as the central atom, there are five *d* valence orbitals in addition to the one *s* and three *p* valence orbitals. Specifically, a first-row transition metal atom has nine valence orbitals—five 3*d* orbitals, one 4*s* orbital, and three 4*p* orbitals.

If the central metal atom used all of its *d*, *s*, and *p* valence orbitals in σ bonding, a total of nine ligands could be attached. However, because of the large size of most ligands it is extremely difficult to achieve a coordination number of nine. Rhenium, a large third-row transition metal, and hydrogen,

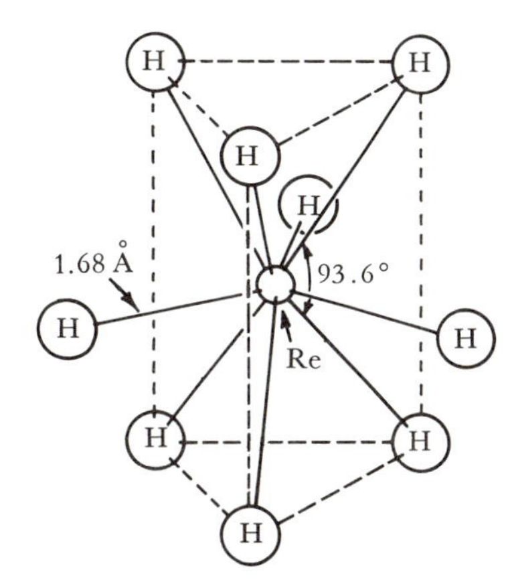

5–2
The structure of the ReH_9^{2-} ion. There are six H atoms at the corners of a trigonal prism, and three more H atoms around the Re atom in a plane halfway between the triangular-end faces of the prism.

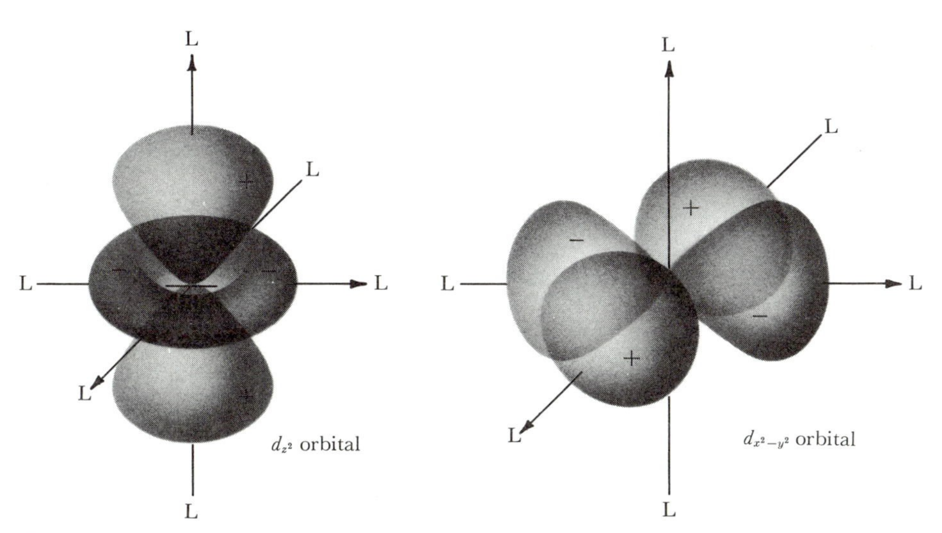

5–3
The spatial orientation of the d_{z^2} and $d_{x^2-y^2}$ orbitals in an octahedral complex. L represents a general ligand.

a small ligand, form the complex ReH_9^{2-}, which exhibits the coordination number of nine. The structure of this complex is shown in Figure 5–2.

Most first-row metal complexes have the coordination number six and an octahedral structure. Several octahedral complexes are shown in Figure 5–1. Six of the nine valence orbitals of the central atom are used in σ bonding in an octahedral structure, namely the $3d_{x^2-y^2}$, $3d_{z^2}$, $4p_x$, $4p_y$, $4p_z$, and $4s$. Notice that the $3d_{x^2-y^2}$ and $3d_{z^2}$ orbitals are directed toward the ligands, as shown in Figure 5–3. In localized molecular-orbital language six d^2sp^3 hybrid orbitals

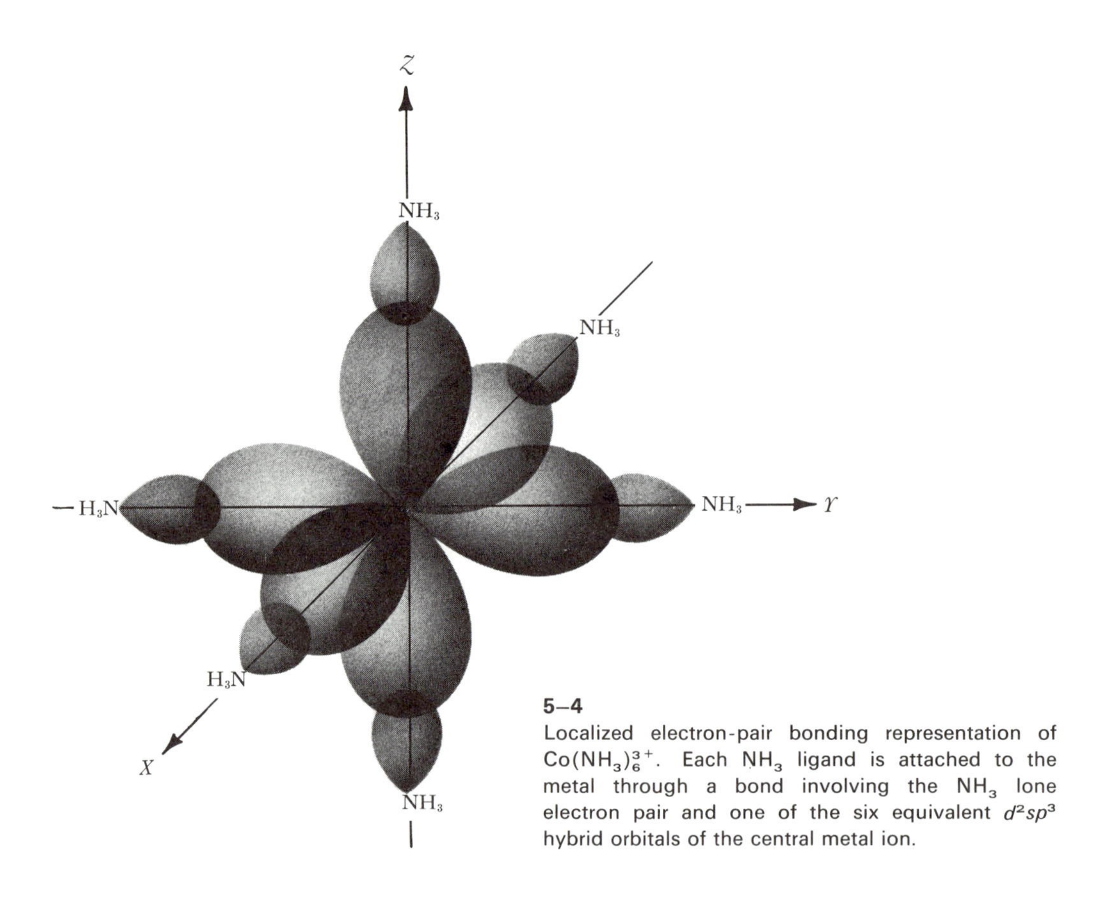

5–4
Localized electron-pair bonding representation of $Co(NH_3)_6^{3+}$. Each NH_3 ligand is attached to the metal through a bond involving the NH_3 lone electron pair and one of the six equivalent d^2sp^3 hybrid orbitals of the central metal ion.

are used to attach the six ligands, as shown in Figure 5–4. However, the localized molecular orbital theory is inadequate for explaining the colors and magnetic properties of transition-metal complexes. The model of greatest utility in discussing the properties of metal complexes is based on delocalized molecular orbital theory. Because the model emphasizes the interaction of the *d* valence orbitals of the metal atoms with appropriate ligand orbitals, it is called *ligand field theory*.

5–2 LIGAND FIELD THEORY FOR OCTAHEDRAL COMPLEXES

Consider the five 3*d* orbitals in a metal complex such as $Co(NH_3)_6^{3+}$. To describe the bonding we choose a coordinate system in which the *X*, *Y*, and *Z* axes go through the nitrogen nuclei. The Co $3d_{x^2-y^2}$ and $3d_{z^2}$ orbitals overlap the approximately sp^3 lone-pair orbitals of the six NH_3 groups, as shown in Figure 5–5. As we did for diatomic molecules, we combine the metal and ligand

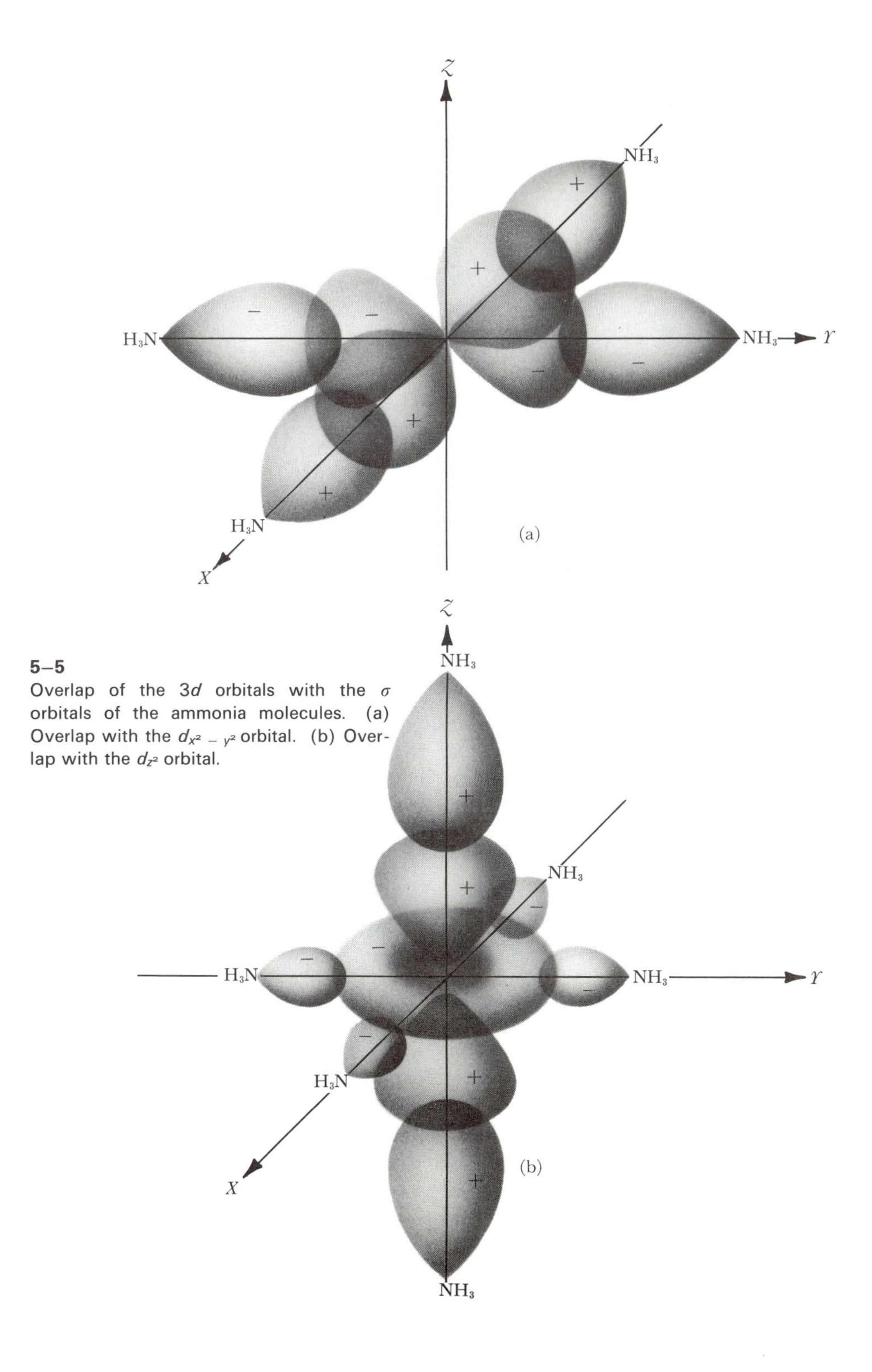

5–5
Overlap of the 3d orbitals with the σ orbitals of the ammonia molecules. (a) Overlap with the $d_{x^2-y^2}$ orbital. (b) Overlap with the d_{z^2} orbital.

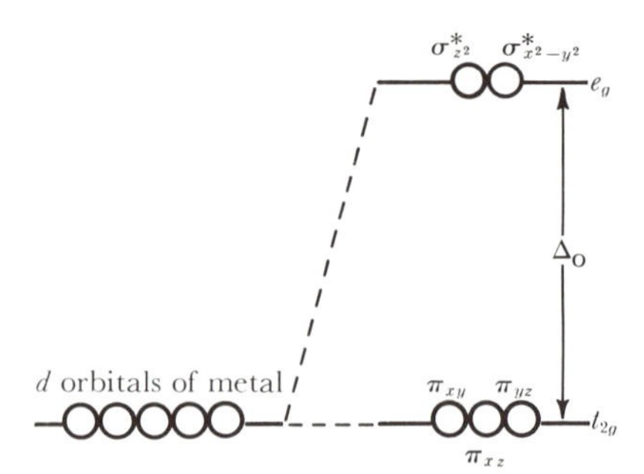

5–6
Ligand-field splitting in an octahedral complex.

orbitals in appropriate linear combinations, thereby forming two bonding orbitals, $\sigma^b_{z^2}$ and $\sigma^b_{x^2-y^2}$, and two antibonding orbitals, $\sigma^*_{z^2}$ and $\sigma^*_{x^2-y^2}$. The antibonding molecular orbitals $\sigma^*_{z^2}$ and $\sigma^*_{x^2-y^2}$ have much more metal than ligand character, and often are referred to simply as the "metal" $3d_{z^2}$ and $3d_{x^2-y^2}$ orbitals.

To construct the σ molecular orbitals we did not use three of the five $3d$ valence orbitals in the cobalt atom, specifically, the $3d_{xz}$, $3d_{yz}$, and $3d_{xy}$. These orbitals are directed between the ligands and are projected into "free" space. Therefore these d orbitals are not situated properly for σ bonding in $Co(NH_3)_6^{3+}$. Consequently the originally degenerate five $3d$ valence orbitals, in which the electrons have equal energy in the free atom, divide into two sets in an octahedral complex. The $3d_{x^2-y^2}$ and $3d_{z^2}$ orbitals are involved in the σ molecular-orbital system, whereas the $3d_{xz}$, $3d_{yz}$, and $3d_{xy}$ orbitals are nonbonding or, with appropriate ligands, may be involved in a π molecular-orbital system.

The two orbital sets, $\sigma^*_{x^2-y^2}$, $\sigma^*_{z^2}$ (or $3d_{x^2-y^2}$, $3d_{z^2}$) and π_{xz}, π_{yz}, π_{xy} (or $3d_{xz}$, $3d_{yz}$, $3d_{xy}$) are called *ligand-field levels*. As shown in Figure 5–6, the $\sigma^*_{x^2-y^2}$ and $\sigma^*_{z^2}$ orbitals always have higher energy than the π_{xz}, π_{yz}, and π_{xy} orbitals in an octahedral complex. The energy separation of the $\sigma^*(d)$ and $\pi(d)$ levels is called the *octahedral ligand-field splitting* and is abbreviated Δ_o. The higher two orbitals commonly are denoted by e_g, and the lower three orbitals are denoted by t_{2g}.

The twelve electrons furnished by the six NH_3 ligands occupy six bonding molecular orbitals constructed from the NH_3 lone-pair orbitals and the $3d_{z^2}$, $3d_{x^2-y^2}$, $4s$, $4p_x$, $4p_y$, and $4p_z$ metal orbitals. The bonding combinations are very stable and will not concern us further. They are analogous to the six d^2sp^3 bonding orbitals in the localized molecular-orbital description. Also, we will not need to consider the antibonding orbitals derived from the $4s$ and $4p$ metal atomic orbitals, because these orbitals are very energetic. The important orbitals for our discussion comprise the two ligand-field levels, t_{2g} and e_g, shown in Figure 5–6. Because the valence electronic structure of Co^{3+} is $3d^6$, there are six valence electrons available to place in the t_{2g} and e_g levels. There are two possibilities, depending on the value of Δ_o. In $Co(NH_3)_6^{3+}$, Δ_o is sufficiently large to allow all six *d* electrons to fill the t_{2g} level, thereby giving the ground-state electronic structure $(t_{2g})^6$, with all electrons paired. However, if Δ_o is smaller than the required electron-pairing energy, as happens in the Co^{3+} complex CoF_6^{3-}, the *d* electrons occupy the t_{2g} and e_g levels to give the maximum number of unpaired spins. The ground-state structure of CoF_6^{3-} is $(t_{2g})^4(e_g)^2$, with four unpaired electrons. Because of the difference in the number of unpaired electron spins in the two complexes, $Co(NH_3)_6^{3+}$ is said to be *low-spin*, whereas CoF_6^{3-} is referred to as a *high-spin complex*.

Table 5–1. Unpaired electrons in uncomplexed metal ions

Ion	Number of *d* electrons	Number of unpaired *d* electrons
Sc^{3+}	0	0
Ti^{3+}	1	1
V^{2+}	3	3
V^{3+}	2	2
Cr^{2+}	4	4
Cr^{3+}	3	3
Mn^{2+}	5	5
Fe^{2+}	6	4
Fe^{3+}	5	5
Co^{2+}	7	3
Co^{3+}	6	4
Ni^{2+}	8	2
Cu^{2+}	9	1
Zn^{2+}	10	0

Next we will consider the ground-state electronic configurations of octahedral complexes that contain central metal ions other than Co^{3+}. Table 5–1 gives the number of unpaired *d* electrons in first-row dipositive and tripositive transition-metal ions that commonly are observed in octahedral coordination complexes. In each case the d^n configuration and the number of unpaired *d* electrons for the uncomplexed metal ion are given for reference. Referring to Figure 5–6 we see that metal ions with one, two, and three valence electrons will have the respective ground-state configurations $(t_{2g})^1$, $(t_{2g})^2$, and $(t_{2g})^3$. There are two possible ground-state configurations for the metal d^4 configuration, depending on the value of Δ_o in the complex. If Δ_o is less than the energy required to pair two *d* electrons in the t_{2g} level, the fourth electron will go into the e_g level, thereby giving the high-spin configuration $(t_{2g})^3(e_g)^1$, with four unpaired electrons. Ligands that form high-spin complexes are called *weak-field ligands*.

However, if Δ_o is larger than the required electron-pairing energy, the fourth electron will go into the lower-energy t_{2g} level and pair with one of the three electrons already present in that level. In this situation the ground state of the complex is the low-spin configuration $(t_{2g})^4$, with only two unpaired electrons. Ligands that cause splittings large enough to allow electrons to occupy preferentially the more stable t_{2g} level to give low-spin complexes are called *strong-field ligands*.

In filling the t_{2g} and e_g energy levels the electronic configurations d^5, d^6, and d^7, as well as d^4, can exhibit either a high-spin or a low-spin ground state, depending on the value of Δ_o in the complex. For a given d^n configuration the paramagnetism of a high-spin complex is larger than that of a low-spin complex. Examples of octahedral complexes with the possible $(t_{2g})^x(e_g)^y$ configurations are given in Table 5–2.

Table 5–2. Electronic configurations of octahedral complexes

Electronic configuration of the metal ion	Electronic structure of the complex	Number of unpaired electrons	Example
$3d^1$	$(t_{2g})^1$	1	$Ti(H_2O)_6^{3+}$
$3d^2$	$(t_{2g})^2$	2	$V(H_2O)_6^{3+}$
$3d^3$	$(t_{2g})^3$	3	$Cr(H_2O)_6^{3+}$
$3d^4$	Low-spin; $(t_{2g})^4$	2	$Mn(CN)_6^{3-}$
	High-spin; $(t_{2g})^3(e_g)^1$	4	$Cr(H_2O)_6^{2+}$
$3d^5$	Low-spin; $(t_{2g})^5$	1	$Fe(CN)_6^{3-}$
	High-spin; $(t_{2g})^3(e_g)^2$	5	$Mn(H_2O)_6^{2+}$
$3d^6$	Low-spin; $(t_{2g})^6$	0	$Co(NH_3)_6^{3+}$
	High-spin; $(t_{2g})^4(e_g)^2$	4	CoF_6^{3-}
$3d^7$	Low-spin; $(t_{2g})^6(e_g)^1$	1	$Co(NO_2)_6^{4-}$
	High-spin; $(t_{2g})^5(e_g)^2$	3	$Co(H_2O)_6^{2+}$
$3d^8$	$(t_{2g})^6(e_g)^2$	2	$Ni(NH_3)_6^{2+}$
$3d^9$	$(t_{2g})^6(e_g)^3$	1	$Cu(H_2O)_6^{2+}$

The first-row transition-metal ions that form the largest number of stable octahedral complexes are Cr^{3+} (d^3), Ni^{2+} (d^8), and Co^{3+} (d^6). The +3 central-ion charge for chromium and cobalt apparently is large enough for strong σ bonding, but not so large that the ligands would be oxidized and the complex destroyed. Furthermore, the orbital configurations $(t_{2g})^3$ and $(t_{2g})^6$ take maximum advantage of the low-energy t_{2g} level; $(t_{2g})^3$ corresponds to a half-filled t_{2g} shell, which requires no electron-pairing energy, and $(t_{2g})^6$ is the closed-shell structure. The $(t_{2g})^6(e_g)^2$ configuration of octahedral Ni^{2+} complexes features a filled t_{2g} and a half-filled e_g level, therefore complexes that exhibit relatively small Δ_o values are quite stable.

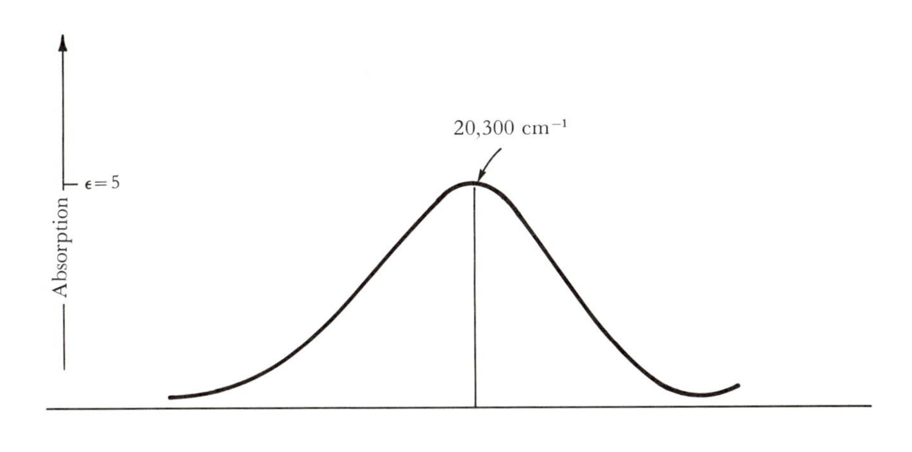

5–7
The absorption spectrum of $Ti(H_2O)_6^{3+}$ in the visible region.

d–d Transitions and light absorption

An electron in the t_{2g} level of an octahedral complex can absorb a light photon in the near-infrared, visible, or ultraviolet region and make a transition to an unoccupied orbital in the more energetic e_g level. Electron excitations of this type are called *d–d transitions*. Although *d–d* transitions generally give rise to weak absorption bands, they are responsible for the characteristic colors of many transition-metal complexes. An example is the red-violet Ti^{3+} complex, $Ti(H_2O)_6^{3+}$, whose ground state is $(t_{2g})^1$. Excitation of the electron from the t_{2g} orbital to the e_g orbital occurs with light absorption in the vicinity of 5000 Å (20,000 cm^{-1}). Figure 5–7 shows that maximum absorption occurs at 4930 Å (20,300 cm^{-1}), with a molar extinction coefficient (ε) of 5. The value of the octahedral ligand-field splitting, Δ_o, usually is expressed in wave numbers; thus we say that for $Ti(H_2O)_6^{3+}$, $\Delta_o = 20{,}300$ cm^{-1}.

The colors of many other transition metal complexes are due to *d–d* transitions. The number of absorption bands depends on the molecular geometry of the complex and the d^n configuration of the central metal atom. The intensities of the *d–d* bands also vary with geometry. The absorption bands in octahedral complexes such as $Ti(H_2O)_6^{3+}$ are weak (ε range of 1–500) because $d-d$ transitions of the type $t_{2g} \rightarrow e_g$ are orbitally forbidden. However, certain of the *d–d* transitions in tetrahedral complexes are fully allowed and can lead to fairly strong absorption (ε range of 200–5000). Thus, in many cases, the study of the electronic spectra of metal complexes is a powerful tool for determining molecular geometries.

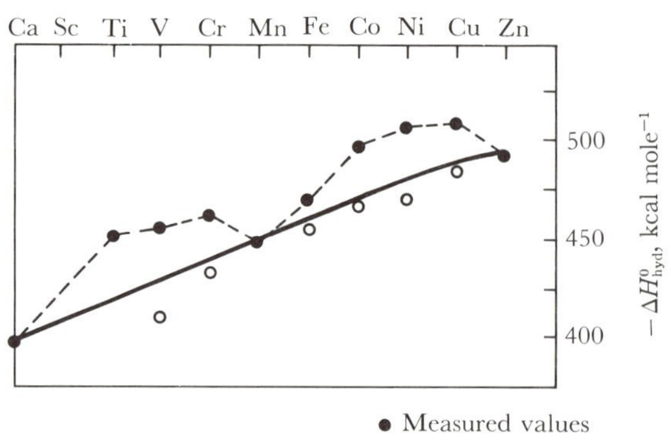

5–8
Measured and ligand-field corrected values of heats of hydration of dipositive transition-metal ions. The heat of hydration is a measure of the $M-OH_2$ bond strength.

A special case of interest is the high-spin d^5 configuration, exhibited by the octahedral complex $Mn(H_2O)_6^{2+}$. The high-spin ground state for the five $3d$ electrons in $Mn(H_2O)_6^{2+}$ is $(t_{2g})^3(e_g)^2$, with five unpaired electrons. All d–d transitions from this ground state are spin-forbidden (the transitions of lowest energy are to excited states with three unpaired electrons). The color of the Mn^{2+} ion in aqueous solution is very pale pink. The color results from the extremely low intensities (ε values of approximately 0.01) of the spin-forbidden d–d absorption bands of $Mn(H_2O)_6^{2+}$, which lie in the visible region. (Weak absorption bands arising from electronic transitions that are "forbidden" according to orbital- or spin-selection rules were discussed first in Section 4–8.)

Heats of hydration of hexaaquo complexes

A direct experimental indication of the special stability of certain d^n configurations in an octahedral ligand field is provided by the observed heats of hydration of dipositive transition-metal ions, as shown in Figure 5–8. If ionic size (thus the charge density on the central metal ion) were the only factor affecting bond energy, the curve would be approximately a straight line, thereby corresponding to a steady increase in bond energy from Ca^{2+} to Zn^{2+} as the ionic radii decrease. The double-humped curve indicates that, except for d^5 Mn^{2+}, the relative stabilization afforded by placing electrons preferentially in the t_{2g} level must make a substantial contribution to the hydration energy. As shown in Figure 5–9, each electron in the t_{2g} level will represent $0.4\Delta_o$ lower energy than the reference zero of energy, whereas each electron in the e_g level will be $0.6\Delta_o$ higher.

The relative ligand-field stabilization energies (LFSE) of the high-spin $M(H_2O)_6^{2+}$ complexes are given in Table 5–3. Ions such as Ti^{2+}, V^{2+}, Co^{2+}, and Ni^{2+} have large LFSE values and correspondingly large negative heats of hydration. After adjusting for LFSE, using Δ_o values derived from d–d

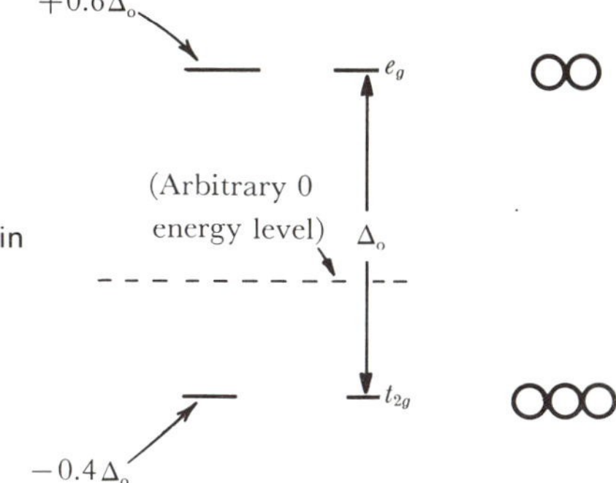

5–9
Relative energies of the t_{2g} and e_g orbitals in an octahedral ligand field.

absorption spectra of the $M(H_2O)_6^{2+}$ complexes, the "corrected" points fall approximately on the Ca^{2+}–Zn^{2+} line (Figure 5–8). The close correlation between data on heats of hydration and *d–d* spectral information demonstrates the power of the ligand field theory.

Table 5–3. Ligand-field stabilization energies of $M(H_2O)_6^{2+}$ complexes

M	Number of *d* electrons Lower level (t_{2g})	Upper level (e_g)	Stabilization energy (Δ_o units)
Ca^{2+}	0	0	0
Ti^{2+}	2	0	0.8
V^{2+}	3	0	1.2
Cr^{2+}	3	1	0.6
Mn^{2+}	3	2	0
Fe^{2+}	4	2	0.4
Co^{2+}	5	2	0.8
Ni^{2+}	6	2	1.2
Cu^{2+}	6	3	0.6
Zn^{2+}	6	4	0

π Bonding in metal complexes

The metal d_{xz}, d_{yz}, and d_{xy} valence orbitals may be used for π bonding to certain types of ligands in octahedral complexes. For example, consider a complex containing six chloride ligands. Each of the metal $d(\pi)$ orbitals overlaps with four ligand $3p(\pi)$ orbitals, as shown in Figure 5–10. In the bonding orbital some electronic charge from the chloride ions is transferred to the metal atom.

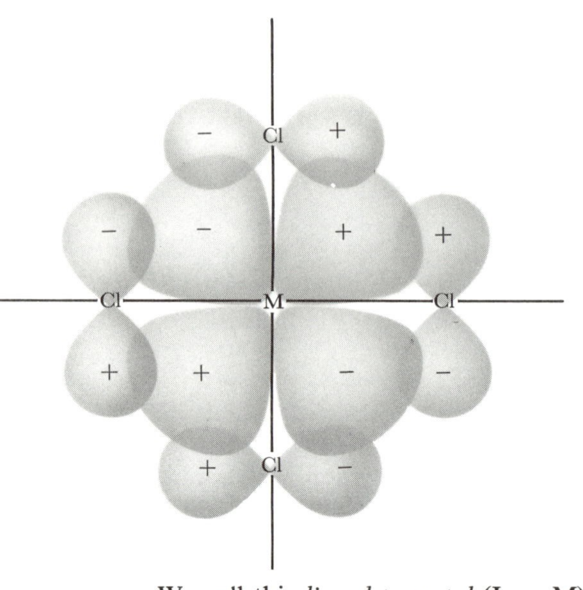

5–10
Overlap of a metal $d(\pi)$ orbital with four ligand $p(\pi)$ orbitals.

We call this *ligand-to-metal* (L → M) π *bonding*. The π orbitals of the metal are destabilized in the process and become antibonding.

If the complex contains a diatomic ligand such as CN^-, two types of π bonding are possible. The occupied π^b ligand orbitals can enter into (L → M) π bonding with the metal $3d_{xz}$, $3d_{yz}$, and $3d_{xy}$ orbitals. However, electrons in the metal t_{2g} level also can be delocalized into the available ligand π^* (CN^-) orbitals, thereby preventing the accumulation of excess negative charge on the metal. This type of bonding removes electron density from the metal and is called *metal-to-ligand* (M → L) π *bonding*. It also commonly is called *back donation or back bonding*. Back donation stabilizes the t_{2g} level and makes it less antibonding. (L → M) π bonding is common when the central metal ion has a large positive charge and empty t_{2g} orbitals. (M → L) π bonding is common when the central metal has low ionic charge and filled t_{2g} orbitals. Both types of π bonding between a $d(\pi)$ orbital and CN^- are shown in Figure 5–11.

5–11▶
Two types of π bonding in metal complexes with a cyanide ion. (a) In the CN^- ion the bonding π^b molecular orbital contains an electron pair and the antibonding π^* orbital (b) is empty. (c) The metal orbitals of the t_{2g} type are more stable in the presence of simple σ-bonding ligands because the t_{2g} orbitals do not concentrate their electrons in the directions of the ligands. But if the ligand has filled π orbitals, then these orbitals interact with the metal t_{2g} orbitals and make them less stable. The splitting constant Δ_o decreases. (d) If the metal has filled t_{2g} orbitals that interact with the empty antibonding π ligand orbitals, then the metal electrons are delocalized, the energy of the orbitals decreases, and the splitting energy, Δ_o, increases. This last effect predominates in most CN^- complexes, thereby producing a large ligand-field splitting.

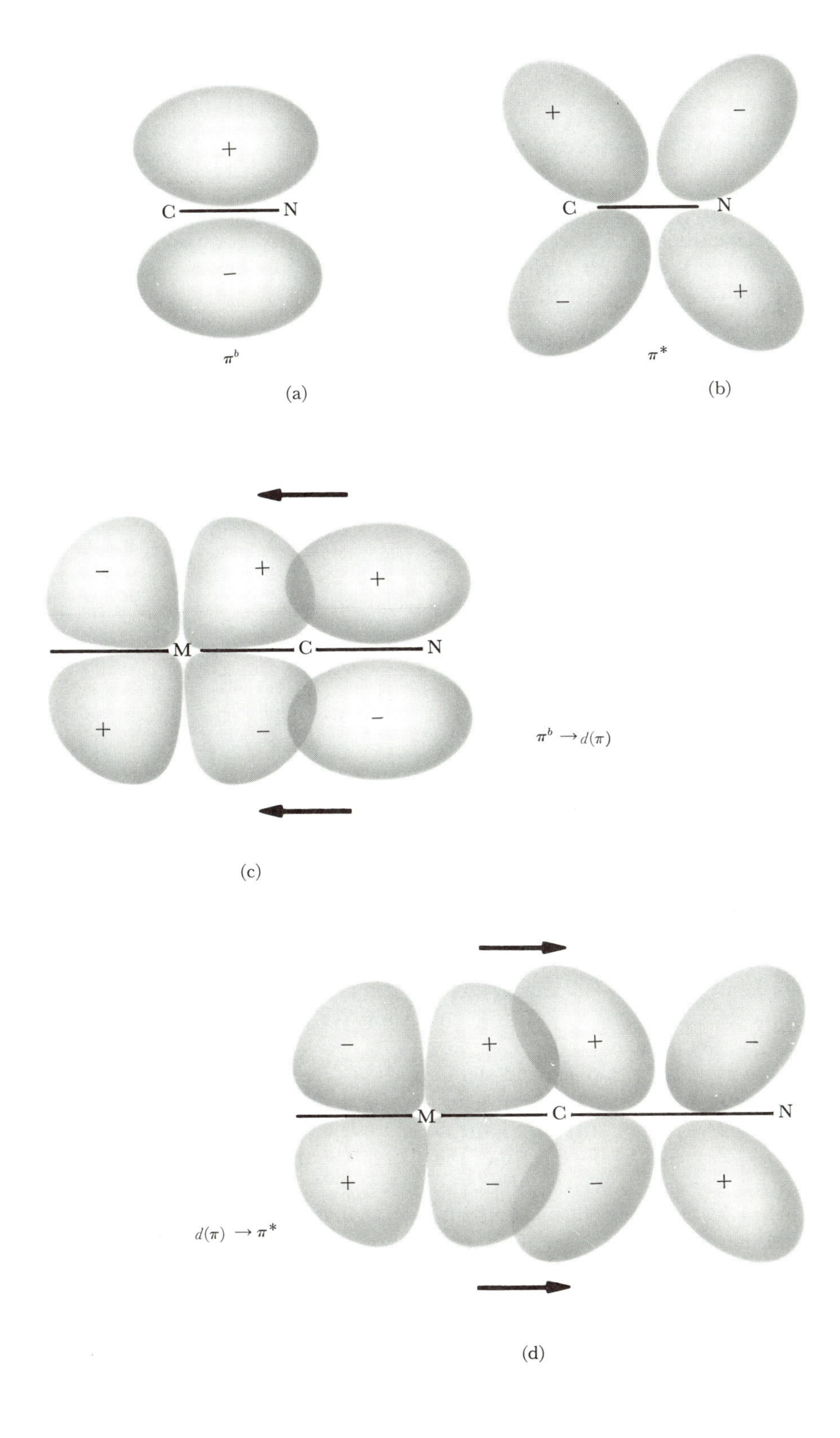
+
C
N
−
π^b
(a)
+
−
C
N
−
+
π^*
(b)
−
+
+
M
C
N
+
−
−
$\pi^b \rightarrow d(\pi)$
(c)
−
+
+
−
M
C
N
+
−
−
+
$d(\pi) \rightarrow \pi^*$
(d)

Factors that influence the value of Δ_o

The Δ_o values for a representative selection of octahedral complexes are given in Table 5–4. The value of Δ_o depends on a number of variables, the most important being the nature of the ligand, the ionic charge (or oxidation number) of the central metal ion, and the principal quantum number, n, of the d valence orbitals. We will discuss these variables individually.

Table 5–4. Values of Δ_o for representative octahedral metal complexes

Octahedral complexes	Δ_o, cm^{-1}	Octahedral complexes	Δ_o, cm^{-1}
TiF_6^{3-}	17,000	$Co(NH_3)_6^{3+}$	22,900
$Ti(H_2O)_6^{3+}$	20,300	$Co(CN)_6^{3-}$	34,500
$V(H_2O)_6^{3+}$	17,850	$Co(H_2O)_6^{2+}$	9,300
$V(H_2O)_6^{2+}$	12,400	$Ni(H_2O)_6^{2+}$	8,500
$Cr(H_2O)_6^{3+}$	17,400	$Ni(NH_3)_6^{2+}$	10,800
$Cr(NH_3)_6^{3+}$	21,600	$RhBr_6^{3-}$	21,600
$Cr(CN)_6^{3-}$	26,600	$RhCl_6^{3-}$	22,800
$Cr(CO)_6$	32,200	$Rh(NH_3)_6^{3+}$	34,100
$Fe(CN)_6^{3-}$	35,000	$Rh(CN)_6^{3-}$	44,000
$Fe(CN)_6^{4-}$	33,800	$IrCl_6^{3-}$	27,600
$Co(H_2O)_6^{3+}$	18,200	$Ir(NH_3)_6^{3+}$	40,000

Nature of the ligand. The ordering of ligands in terms of their ability to split the e_g and t_{2g} molecular orbitals is known as the *spectrochemical series.* The order of ligand-field splitting of some important ligands is

$$CO, CN^- > NO_2^- > NH_3 > OH_2 > OH^- > F^- > -SCN^-, Cl^- > Br^- > I^-$$

Octahedral complexes containing ligands such as CN^- and CO, which are at the strong-field end of the spectrochemical series, have Δ_o values between 30,000 cm^{-1} and 50,000 cm^{-1}. At the other end of the series, octahedral complexes containing Br^- and I^- have relatively small Δ_o values, in many cases less than 20,000 cm^{-1}.

We already have discussed the important types of metal–ligand bonding in transition-metal complexes. The manner in which each type affects the value of Δ_o is illustrated in Figure 5–12. We see that a strong ($L \rightarrow M$) σ interaction increases the energy of the e_g orbitals, thereby increasing the value of Δ_o. A strong ($L \rightarrow M$) π interaction destabilizes t_{2g}, thereby decreasing the value of Δ_o. A strong ($M \rightarrow L$) π^* interaction lowers the energy of t_{2g}, thereby increasing the value of Δ_o. The spectrochemical series correlates reasonably well with the relative π-donor and π-acceptor abilities of the ligands. The π-acceptor ligands [those capable of strong ($M \rightarrow L$) π^* bonding] cause large

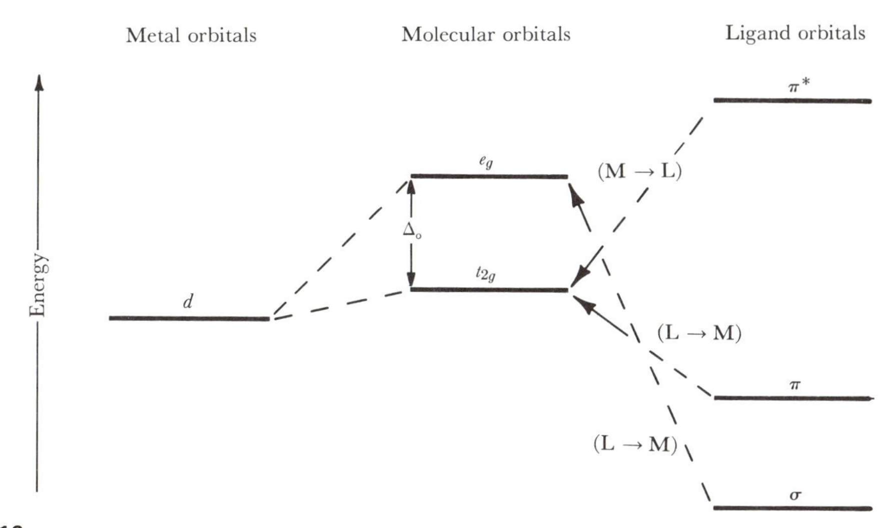

5-12
The effect on the value of Δ_o of interaction of the ligand σ, π, and π^* orbitals with the metal *d* orbitals.

splittings, whereas the π-donor ligands [those capable of strong (L → M) π donation] cause small splittings. The ligands with intermediate Δ_o values have little or no π-bonding capabilities.

The correlation of the spectrochemical series with the π-donor and π-acceptor characteristics of several important ligands is summarized as follows:

$CO, CN^- > NO_2^- >$	$NH_3 > OH_2 >$	$OH^- > F^- >$	$-SCN^-, Cl^- > Br^- > I^-$
π acceptors	non-π bonding	weak π donors	π donors

Ionic charge of the central metal ion. In complexes containing ligands that are neither strong π donors nor π acceptors, Δ_o increases with increasing ionic charge on the central metal ion. An example is the increase in Δ_o from $V(H_2O)_6^{2+}$, with $\Delta_o = 12{,}400\ cm^{-1}$, to $V(H_2O)_6^{3+}$, with $\Delta_o = 17{,}850\ cm^{-1}$. The larger Δ_o of $V(H_2O)_6^{3+}$ probably is due to a substantial increase in σ bonding of the H_2O ligands to the more positive V^{3+} central metal ion. The enhanced σ interaction results in an increase in the difference in energy between e_g and t_{2g}.

In octahedral complexes containing strong π-acceptor ligands, an increase in oxidation number of the metal ion does not seem to be accompanied by a substantial increase in Δ_o. For example, both $Fe(CN)_6^{4-}$ and $Fe(CN)_6^{3-}$ have Δ_o values of approximately $34{,}000\ cm^{-1}$. When the charge on the central metal ion is increased from +2 in $Fe(CN)_6^{4-}$ to +3 in $Fe(CN)_6^{3-}$, the t_{2g} level apparently is destabilized through a decrease in π back bonding equally as much as the energy of the e_g level is raised by the greater (L → M) σ bonding.

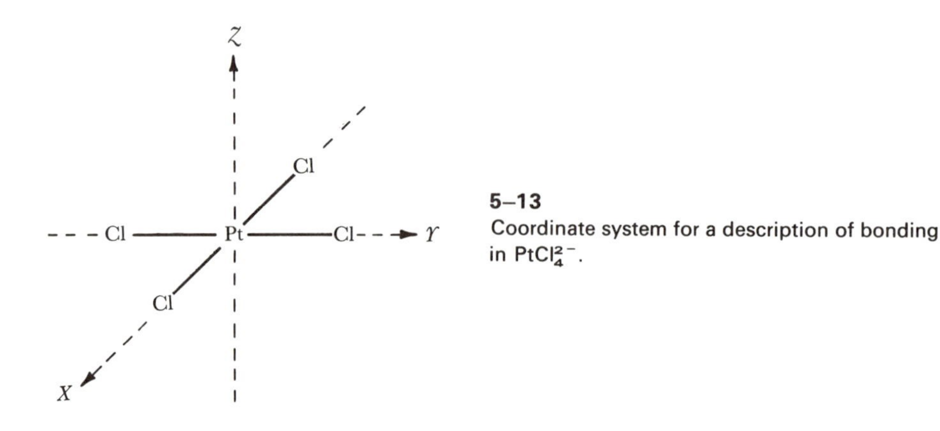

5–13
Coordinate system for a description of bonding in $PtCl_4^{2-}$.

Principal quantum number of the d valence orbitals. For changes from 3*d* to 4*d* to 5*d* valence orbitals of the central metal ion in an analogous series of complexes, the value of Δ_o increases markedly. For example, the Δ_o values for $Co(NH_3)_6^{3+}$, $Rh(NH_3)_6^{3+}$, and $Ir(NH_3)_6^{3+}$ are 22,900 cm^{-1}, 34,100 cm^{-1}, and 40,000 cm^{-1}, respectively. Presumably the 4*d* and 5*d* valence orbitals of the ion are more suitable for σ bonding with the ligands than are the 3*d* orbitals, but the reason for this is not well understood. An important consequence of the much larger Δ_o values of 4*d* and 5*d* central metal ions is that *all* second- and third-row metal complexes have low-spin ground states, even complexes such as $RhBr_6^{3-}$, which contain ligands at the weak-field end of the spectrochemical series.

5–3 LIGAND FIELD THEORY FOR SQUARE PLANAR COMPLEXES

As we mentioned previously, many d^8 central metal ions form square planar complexes. The example we will use here is $PtCl_4^{2-}$, pictured in a reference coordinate system in Figure 5–13. The principal σ bonding involves the overlap of $3p(\sigma)$ Cl^- orbitals with the $5d_{x^2-y^2}$, $6s$, $6p_x$, and $6p_y$ metal valence orbitals. In the language of localized molecular orbital theory, the σ bonding is summarized as dsp^2 in a square planar complex.

Of principal interest here is the ligand-field splitting of the antibonding molecular orbitals derived from the metal *d* valence orbitals in a square planar complex. Examination of the overlaps of the Cl^- 3*p* orbitals with the Pt^{2+} 5*d* valence orbitals in $PtCl_4^{2-}$ (Figure 5–14) reveals that only one *d* orbital is involved in strong σ bonding in a square planar complex, namely the $d_{x^2-y^2}$ orbital. The d_{z^2} orbital interacts weakly with the four ligand σ valence orbitals because most of the d_{z^2} orbital is directed along the *Z* axis away from the ligands. The metal d_{xz}, d_{yz}, and d_{xy} valence orbitals are involved in π molecular orbitals

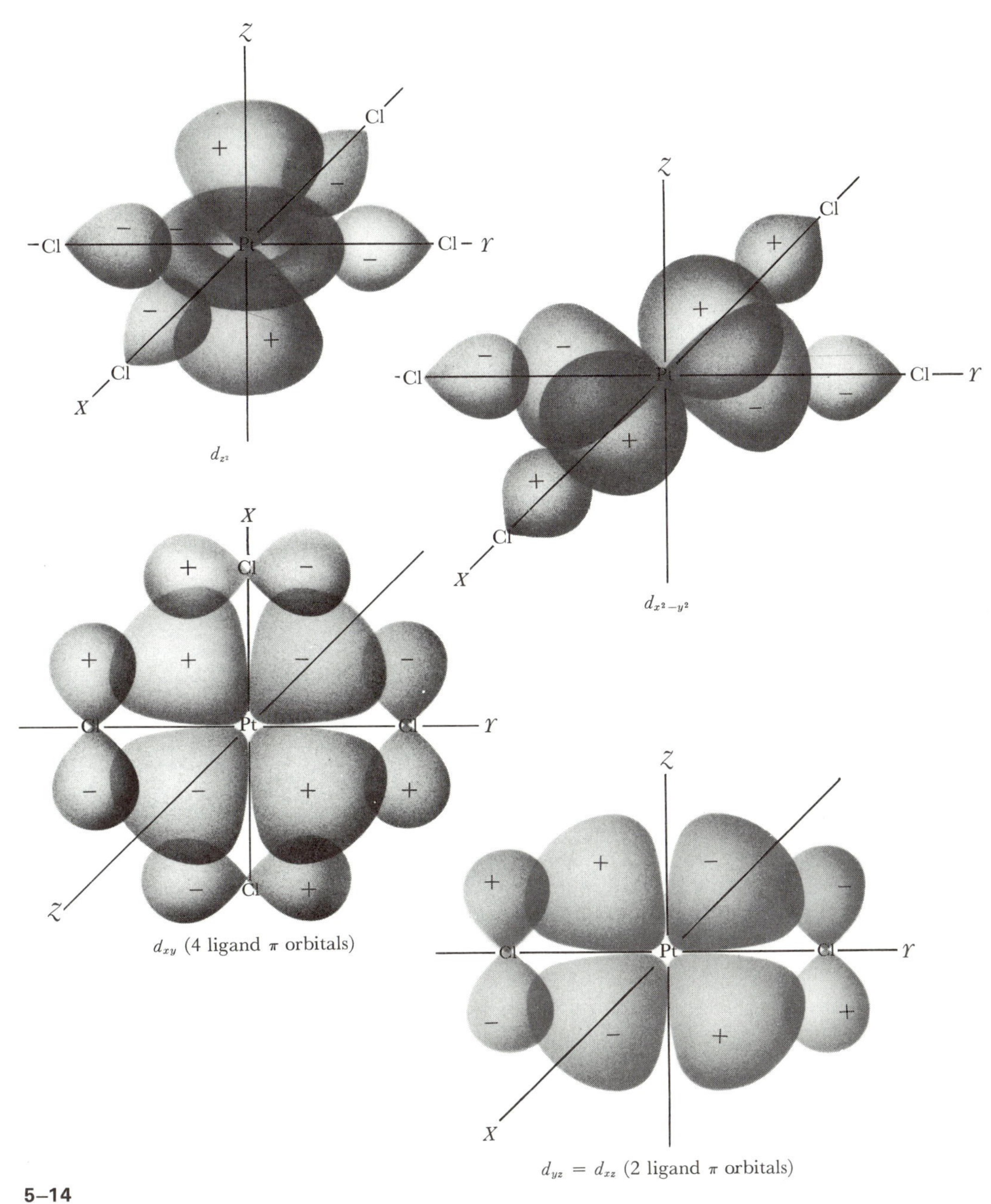

5–14
Overlap of the metal *d* valence orbitals with the Cl^- ligand valence (3*p*) orbitals in a square planar complex.

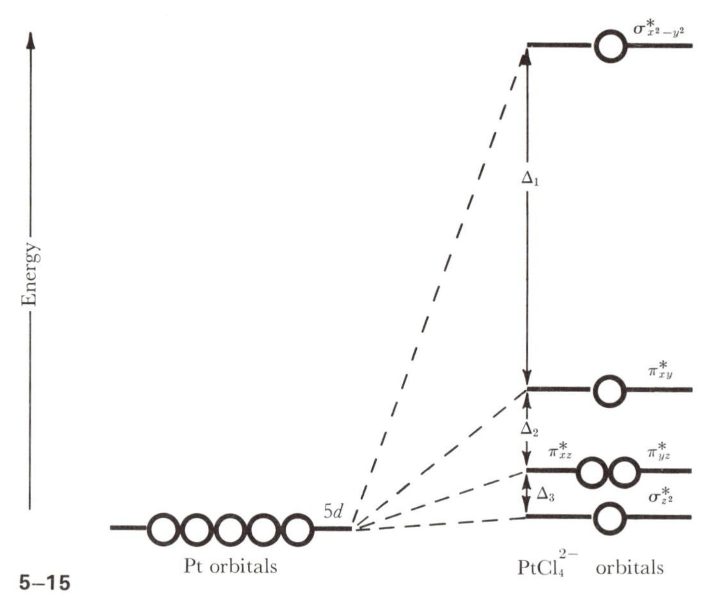

5–15

Ligand-field splitting diagram for the square planar complex $PtCl_4^{2-}$. For certain other square planar complexes, the $\sigma^*_{z^2}$ level may lie above π^*_{xz}, π^*_{yz} in energy.

with the ligands. The d_{xy} orbital interacts with $3p(\pi)$ valence orbitals on all four ligands, whereas each of the equivalent d_{xz} and d_{yz} orbitals interacts with only two ligands.

The ligand-field splitting in a square planar complex is rather complicated because there are four different energy levels. The ligand-field splitting diagram, which has been worked out from spectral studies of $PtCl_4^{2-}$, is shown in Figure 5–15. For all square planar complexes it is reasonable to place the strongly antibonding $\sigma^*_{x^2-y^2}$ orbital highest in energy. We also can position π^*_{xy} above π^*_{xz} (π^*_{yz}), since d_{xy} interacts with all four ligands. The position of the weakly antibonding $\sigma^*_{z^2}$ level probably varies in square planar complexes, depending on the nature of the ligand and the metal. As we have indicated in Figure 5–15, recent studies of $PtCl_4^{2-}$ allow us to position $\sigma^*_{z^2}$ below the π^*_{xz}–π^*_{yz} level. However, regardless of the placement of $\sigma^*_{z^2}$, the most important characteristic of the ligand-field splitting in a square planar complex is that $\sigma^*_{x^2-y^2}$ is of much higher energy than the other four orbitals, which are about the same energy.

The valence electronic configuration of the Pt^{2+} ion is $5d^8$. Because the Δ_1 splitting for $PtCl_4^{2-}$ (Figure 5–15) is much larger than the energy required to pair two electrons in this complex, the ground-state electronic structure is $(\sigma^*_{z^2})^2(\pi^*_{xz,yz})^4(\pi^*_{xy})^2$. In agreement with the ligand-field model, experimental studies show that $PtCl_4^{2-}$ is diamagnetic.

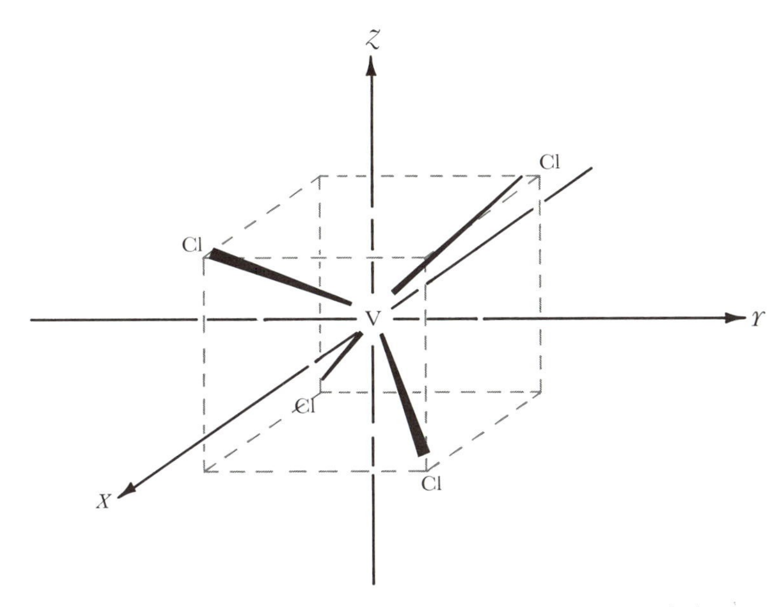

5–16
Coordinate system for tetrahedral VCl_4.

We conclude from the ligand-field splitting (Figure 5–15) that a particularly favorable ground-state electronic structure for square planar complexes is the low-spin configuration $(\sigma^*_{z^2})^2(\pi^*_{xz,yz})^4(\pi^*_{xy})^2$. The four relatively stable orbitals are occupied completely in this arrangement, and the high-energy $\sigma^*_{x^2-y^2}$ orbital is left vacant. Therefore the ligand-field model is consistent with the fact that complexes of the d^8 metal ions, particularly Ni^{2+}, Pd^{2+}, Pt^{2+}, and Au^{3+}, commonly exhibit square planar geometry. All d^8 square planar complexes are known to have low-spin ground-state configurations.

5–4 LIGAND FIELD THEORY FOR TETRAHEDRAL COMPLEXES

An example of a tetrahedral metal complex is VCl_4, which is shown in a convenient coordinate system in Figure 5–16. We discussed the role of s and p valence orbitals in a tetrahedral molecule in Chapter 4. The $4s$ and $4p$ atomic orbitals of vanadium can be used to form σ molecular orbitals. Although the overlap patterns are rather complicated, the $3d_{xz}$, $3d_{yz}$, and $3d_{xy}$ valence orbitals also are situated properly to form σ molecular orbitals. In terms of localized molecular orbitals both sd^3 and sp^3 hybrid orbitals are tetrahedrally oriented. The $3d_{x^2-y^2}$ and $3d_{z^2}$ orbitals of the central atom interact very weakly with the ligands to form π molecular orbitals.

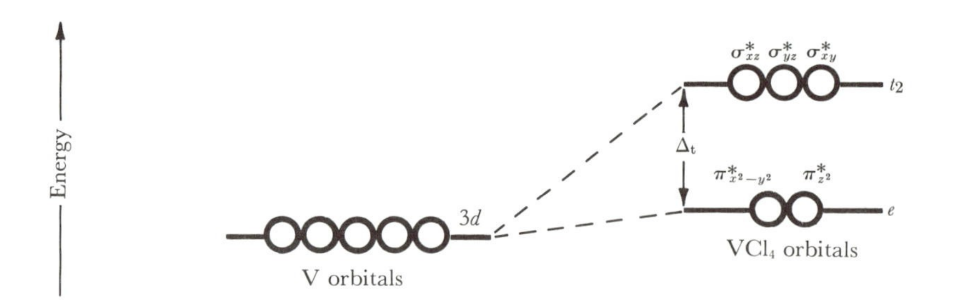

5–17
Ligand-field splitting in a tetrahedral complex. The antibonding orbitals are divided into two sets: (1) three $\sigma^*(d)$ orbitals—the t_2 set; and (2) two $\pi^*(d)$ orbitals—the e set.

The ligand-field splitting diagram for a tetrahedral complex such as VCl_4 is shown in Figure 5–17. The antibonding molecular orbitals derived from the $3d$ valence orbitals are divided into two sets. The orbitals formed from the $3d_{xz}$, $3d_{yz}$, and $3d_{xy}$ orbitals are of higher energy than those formed from the $3d_{z^2}$ and $3d_{x^2-y^2}$ orbitals. Thus the change from octahedral to tetrahedral geometry exactly reverses the role and the energies of the d valence orbitals of the central metal ion. In a tetrahedral complex we call the three $\sigma^*(d)$ orbitals the t_2 set, and the two $\pi^*(d)$ orbitals the e set. We designate the difference in energy between t_2 and e in a tetrahedral complex as Δ_t.

With one valence electron from V^{4+} ($3d^1$), the ground-state structure of VCl_4 is $(e)^1$. The paramagnetism of VCl_4 is consistent with this configuration, which has one unpaired electron. Energy in the near-infrared region excites the electron in e to t_2, with maximum absorption at 9010 cm^{-1}. Thus for VCl_4, Δ_t is 9010 cm^{-1}. The molar extinction coefficient, ε, of the $e \rightarrow t_2$ band is 130, which is larger than a typical value for an octahedral complex, as was discussed previously.

Table 5–5. Values of Δ_t for representative tetrahedral complexes

Tetrahedral complexes	Δ_t, cm^{-1}
VCl_4	9010
$CoCl_4^{2-}$	3300
$CoBr_4^{2-}$	2900
CoI_4^{2-}	2700
$Co(NCS)_4^{2-}$	4700

Values of Δ_t for several representative tetrahedral complexes are given in Table 5–5. Tetrahedral ligand-field splitting (Δ_t) is much smaller than octahedral splitting (Δ_o) for a given metal ion and ligand, and the relationship $\Delta_t \cong 0.45\Delta_o$ has been established from both theoretical and experimental studies. From ligand field theory we predict that the t_2 orbitals in a tetrahedral complex will not form as strong σ bonds with ligand σ orbitals as will the e_g octahedral orbitals, thereby resulting in a much less energetic t_2 level and a relatively small Δ_t value. Because of the small Δ_t values, *all* tetrahedral transition-metal complexes have high-spin ground-state configurations.

Of the dipositive metal ions, Co^{2+} (d^7) is exceptionally stable in tetrahedral complexes. Examples are $CoCl_4^{2-}$, $Co(NCS)_4^{2-}$, and $Co(OH)_4^{2-}$. The relative stability of tetrahedral Co^{2+} complexes is consistent with the fact that the $(e)^4(t_2)^3$ configuration makes maximum use of the lower-energy e level and therefore has a favorable LFSE.

5–5 CHARGE-TRANSFER ABSORPTION BANDS

For many complexes absorption bands are observed that occur in different positions from those associated with d–d transitions of the central metal ion. These bands usually are in the ultraviolet region, but sometimes they occur in the visible portion of the spectrum. The bands often are quite intense because they generally involve fully allowed transitions. The types of electronic excitation that result in these strong absorptions are illustrated schematically:

$$(M)(L) \xrightarrow{h\nu} (M^+)(L^-)$$

metal-to-ligand (or M → L) charge-transfer transition

$$(M)(L) \xrightarrow{h\nu} (M^-)(L^+)$$

ligand-to-metal (or L → M) charge-transfer transition

Absorptions due to M → L or L → M excitation are called *charge-transfer bands*, because the transition involved requires the transfer of electronic charge to either the ligand (M → L) or the metal (L → M). The energies of charge-transfer bands are related closely to the oxidizing (or reducing) abilities of the metal and the ligands. The powerful oxidizing agent MnO_4^- exhibits strong L → M ($O^{2-} \rightarrow Mn^{7+}$) charge-transfer absorption in the visible region ($\bar{\nu}_{max}$ = 18,000 cm^{-1}), thereby giving the permanganate ion its intense purple color. The related chromate ion, CrO_4^{2-}, is yellow because its lowest-energy L → M ($O^{2-} \rightarrow Cr^{6+}$) band is shifted to a higher wave number ($\bar{\nu}_{max}$ = 26,000 cm^{-1}).

In both MnO_4^- and CrO_4^{2-} the lowest-energy L → M charge-transfer band is believed to be due to the excitation of a nonbonding $2p$ oxygen electron to the unoccupied e ligand-field level of the d^0 tetrahedral complex.

Charge transfer of the M → L type occurs at relatively low energy if the central metal atom or ion has reducing properties, and if the attached ligands have unoccupied orbitals of low enough energy to accept the electron. As one example, the complex $Cr(CO)_6$, which contains Cr(0) and the strong π-acceptor ligand CO, absorbs strongly in the ultraviolet region, with $\bar{\nu}_{max} = 35{,}000\ cm^{-1}$. The absorption has been attributed to electron excitation from the filled t_{2g} level of Cr(0) to the unoccupied π^* level of CO.

5-6 EFFECT OF THE NATURE OF THE METAL AND THE LIGAND ON THE STABILITIES OF COMPLEXES

The reaction between a ligand and a metal ion often is classified as a *Lewis acid-base interaction*, in which the ligand (base) donates an electron pair to the metal ion (acid). Two general classes of central metal ions have been designated, according to their ability to attach various ligands. Class a comprises metal ions that form complexes with ligands in the following order of decreasing stability. Ligands containing the donor atoms N, O, and F form more stable complexes than those containing P, S, and Cl, which are more stable than those containing As, Se, and Br. These in turn are stronger than those containing Sb, Te, and I. Conversely, Class b metal ions form stronger complexes with ligands containing heavier donor atoms than with those containing N, O, and F. For instance, Al^{3+} and Fe^{3+} are Class a ions that form much more stable complexes with F^- and OH^- than with Cl^- and HS^- (or S^{2-}). An example of a Class b ion is Hg^{2+}, which forms the very weak complex HgF^+ and the increasingly more stable complexes $HgCl_4^{2-}$, $HgBr_4^{2-}$, and HgI_4^{2-}. Class b ions also tend to complex more strongly with NH_3 than with H_2O, and more strongly with H_2O than with HF. Thus Ag^+ forms $Ag(NH_3)_2^+$ in aqueous NH_3 solution.

Class a metal ions generally are nonpolarizable and combine most effectively with small, "hard" (or nonpolarizable) ligands. In contrast, Class b ions are polarizable and combine with large, "soft" (or polarizable) ligands. R. G. Pearson, of Northwestern University, has suggested the designations "hard" and "soft" for Class a and Class b acids and bases, respectively. The general rule is that soft acids combine most effectively with soft bases and hard acids combine most effectively with hard bases.

Ions with the electronic structures $4d^{10}5s^2$ or $5d^{10}6s^2$ are intermediate in behavior. When one ligand is bonded to the metal ion it acts like a Class a ion, but when four to six ligands are bonded it acts like a Class b ion. Thus PbF^+ is more stable than $PbCl^+$, but $PbCl_4^{2-}$ forms in concentrated HCl, whereas

Table 5–6. Classification of positive ions (acids) as complex formers

Class a (hard)	Intermediate	Class b (soft)
Be^{2+}	Tl^{+}	Cu^{2+}
B^{3+}	Pb^{2+}	Ag^{+}
Mg^{2+}	Bi^{3+}	Pt^{2+}
Al^{3+}	Sn^{2+}	Pt^{4+}
Ti^{4+}	Sb^{3+}	Pd^{2+}
Fe^{3+}		Pd^{4+}
Mn^{2+}		Ir^{3+}
Zn^{2+}		Rh^{3+}
Ga^{3+}		Hg^{2+}
Si^{4+}		

PbF_4^{2-} is not known to exist in aqueous solution. Most Class a ions have high positive charges, small radii, and closed-shell (Al^{3+}) or half-filled *d* shell (Mn^{2+}, Fe^{3+}) configurations. Most Class b ions have low positive charges, large radii, and nonclosed-shell configurations. Most Class b elements are found at or near the right side of each transition series. Table 5–6 classifies some common metal ions as Class a, Class b, or intermediate

Chelation and stability

Ligands that have two or more donor atoms that are situated so they can bond to the central metal ion often form unusually stable complexes. An example is the organic compound ethylenediamine, $H_2NCH_2CH_2NH_2$, which has two nitrogen atoms with ammonialike structures, each of which is about as basic as ammonia:

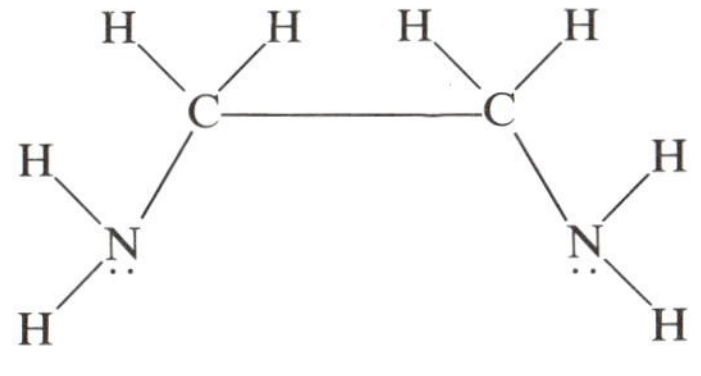

ethylenediamine (en)

Therefore we might expect one ethylenediamine molecule to be equivalent to two ammonia molecules in complexing ability. However, the equilibrium constants for complex formation (K_f values) in Table 5–7 show that ethylenediamine is bonded much more strongly to Ni^{2+} in aqueous solution than are two ammonia molecules.

Complexes with ligands that contain more than one point of attachment are called *chelates*, from the Greek word for crab's claw. Ligands with two

Table 5–7. Standard free energies (ΔG° values) for the formation of Ni^{2+} complexes with NH_3 and $H_2NCH_2CH_2NH_2$ (en) from $Ni(H_2O)_6^{2+}$ at 298°K

Complex	$-\Delta G^\circ$, kcal	K_f
$Ni(NH_3)_2(H_2O)_4^{2+}$	6.6	1.1×10^5
$Ni(en)(H_2O)_4^{2+}$	10.6	4.5×10^7
$Ni(NH_3)_4(H_2O)_2^{2+}$	10.3	1.0×10^8
$Ni(en)_2(H_2O)_2^{2+}$	19.5	1×10^{14}
$Ni(NH_3)_6^{2+}$	11.0	5.5×10^8
$Ni(en)_3^{2+}$	25.7	4.0×10^{18}

points of attachment are called *bidentates*, those with three points are called *tridentates*, and so on. The enhanced stability of chelates, as illustrated by the comparison of $Ni(en)_3^{2+}$ and $Ni(NH)_3)_6^{2+}$, often is called the *chelate effect*. Recent research has led to the discovery of a wide variety of polydentate ligands that wrap around central metal ions, thereby forming extremely stable complexes because of the chelate effect. A chelate containing the important hexadentate ligand EDTA (ethylenediaminetetraacetate) is shown in Figure 5–18.

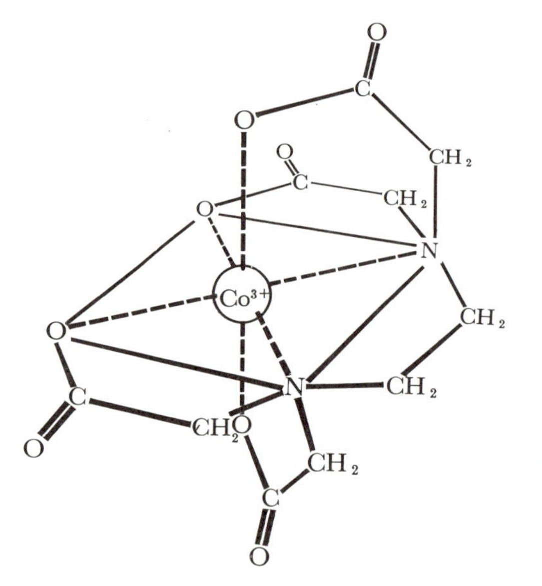

5–18
The hexadentate chelating agent ethylenediaminetetraacetate (EDTA) can occupy all six octahedral coordination positions. The Co^{3+} chelate shown is a uninegative ion and usually is abbreviated $Co(EDTA)^-$. EDTA is such a strong chelating agent that it will remove metals from enzymes and will inhibit their catalytic activity completely.

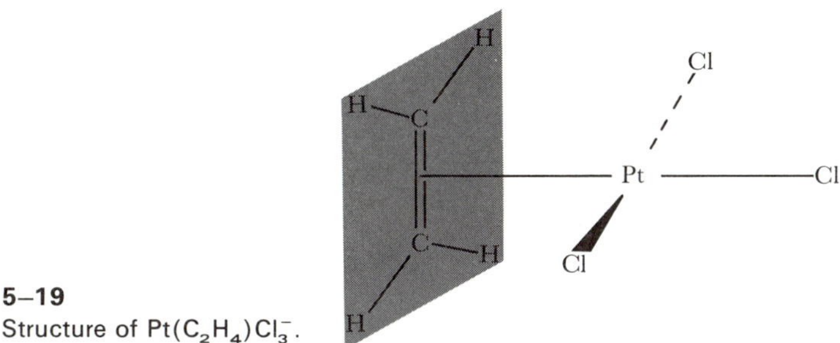

5–19
Structure of $Pt(C_2H_4)Cl_3^-$.

5–7 ORGANOMETALLIC π COMPLEXES OF TRANSITION METALS

Molecules that contain centrally located transition-metal atoms and organic groups attached through delocalized π-orbital networks are called *organometallic π complexes*. The first complexes of this type were discovered by the Danish chemist W. C. Zeise in 1827. The most famous complex is $K[Pt(C_2H_4)Cl_3]$, in which the organic molecule ethylene (C_2H_4) is bound to Pt^{2+}. More than one hundred years after its discovery, the structure of Zeise's salt was determined by x-ray diffraction methods. The structure of the $Pt(C_2H_4)Cl_3^-$ ion is shown in Figure 5–19. The complex can be thought of as having a square planar structure (common to Pt^{2+}) with the ethylene bonded at one corner of the square.

The bonding between C_2H_4 and Pt^{2+} is described conveniently by the model shown in Figure 5–20. The π molecular orbitals of C_2H_4 are used to bond the molecule to the central metal ion. The filled π^b molecular orbital of C_2H_4 is used to form a σ-donor bond with an available Pt^{2+} σ orbital (a combination of the $d_{x^2-y^2}$, s, and p_x orbitals in the model shown). In addition, there is the possibility of forming a π bond between the filled metal d_{xz} orbital and the empty π^* orbital of C_2H_4. This π back bond prevents accumulation of electron density on the metal.

Recent interest in metal π complexes can be traced to 1948, when S. A. Miller and his colleagues at the British Oxygen Company discovered that the organic compound cyclopentadiene (Figure 5–21) reacted with an iron-containing catalyst. The product of this reaction was a stable, orange, crystalline substance. This work was not published until 1951, when T. J. Kealy and P. L. Pauson accidently made the same material by another method. The orange substance has the formula $(C_5H_5)_2Fe$ and is called *ferrocene*.

The structure of ferrocene is like a sandwich in which the meat is the central iron atom and the two bread slices are the cyclopentadienyl groups, as

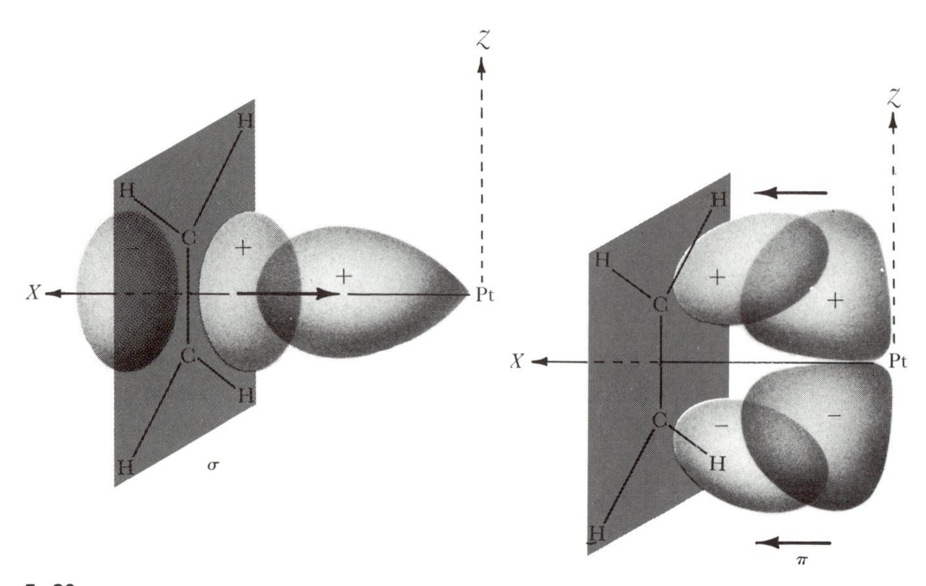

5–20
The π molecular orbitals of C_2H_4 are used to bond the molecule to Pt^{2+} in $Pt(C_2H_4)Cl_3^-$.

shown in Figure 5–22. We can consider the complex as containing the d^6 central ion Fe^{2+} and two coordinated cyclopentadienyl anions ($C_5H_5^-$). The bonding in ferrocene commonly is described in terms of a molecular-orbital model, starting with the delocalized π molecular orbitals of the $C_5H_5^-$ groups. Using the coordinate system shown in Figure 5–22, the d_{xz} and d_{yz} orbitals of Fe^{2+} are situated for strong bonding with the π orbitals of the $C_5H_5^-$ groups. In addition to the d_{xz} and d_{yz} orbitals the s, p_x, p_y, and p_z iron valence orbitals presumably can be used in strong bonding with the $C_5H_5^-$ groups. (The other d orbitals, d_{z^2}, $d_{x^2-y^2}$, and d_{xy}, play a much smaller role in the bonding.) Thus there are six bonding orbitals, which can accommodate twelve electrons. Each $C_5H_5^-$ group furnishes six π electrons, therefore the bonding orbitals are filled in ferrocene.

Recent research has established that the ligand-field splitting in ferrocene is that shown in Figure 5–23. The six electrons furnished by Fe^{2+} (d^6) are accommodated in the relatively low-energy d orbitals, thereby giving the ground-state configuration $(d_{x^2-y^2}, d_{xy})^4(d_{z^2})^2$. Ferrocene is diamagnetic, which agrees with this model.

The complete analysis of the ferrocene complex is as follows. Six electrons are furnished by the iron ion and are accommodated in relatively stable d orbitals. Twelve electrons are furnished by two cyclopentadienyl ions and are accommodated in six stable bonding orbitals, which are constructed from

5–21
Structural formula of cyclopentadiene, C_5H_6.

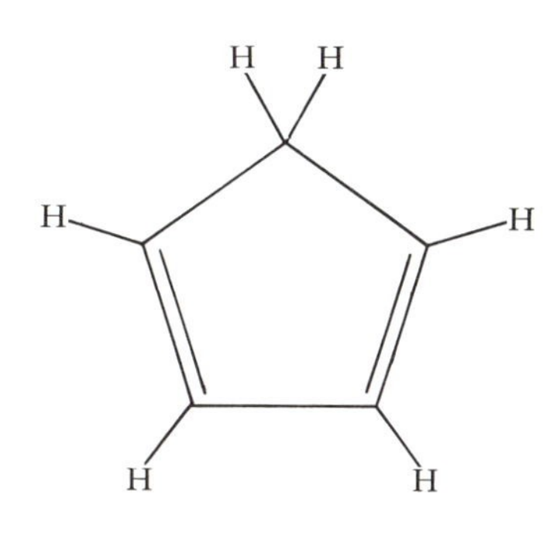

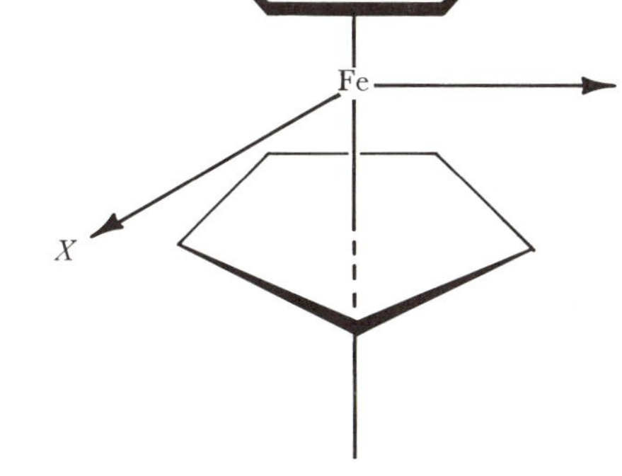

5–22
"Sandwich" structure of ferrocene, $(C_5H_5)_2Fe$.

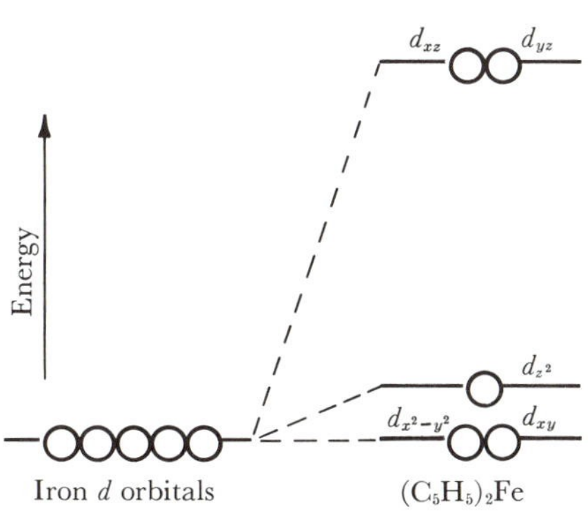

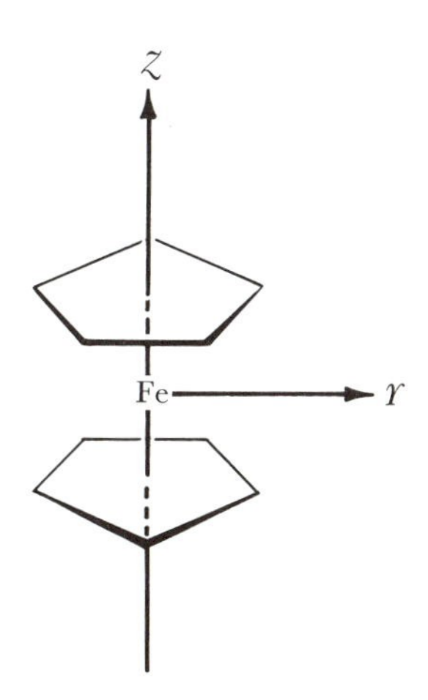

5–23
Ligand-field splitting diagram for ferrocene.

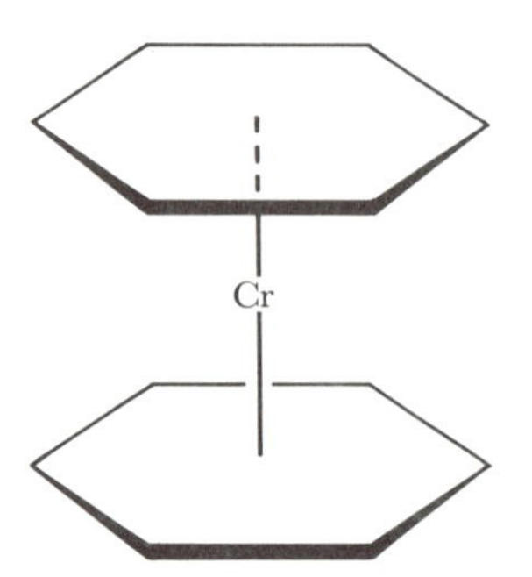

5–24
Structure of dibenzenechromium, $(C_6H_6)_2Cr$.

the $3d_{xz}$, $3d_{yz}$, $4s$, $4p_x$, $4p_y$, and $4p_z$ Fe^{2+} orbitals and the occupied π molecular orbitals of the $C_5H_5^-$ ions. Using this type of theoretical analysis several chemists reasoned (correctly) that many other sandwich complexes, in which the ligands furnish a total of twelve electrons and the metal atom furnishes six electrons, would be stable. The existence of a stable complex containing two molecules of benzene (six π electrons each) and a central chromium atom [Cr(0); six metal valence electrons] is one noteworthy example. This sandwich complex is called *dibenzenechromium* and has the formula $(C_6H_6)_2Cr$. A schematic drawing of the structure of dibenzenechromium is shown in Figure 5–24.

Zeise's anion, ferrocene, and dibenzenechromium are only three examples of organometallic π complexes. In the past twenty years many thousands of such complexes have been prepared and studied, and organometallic chemistry is now a major area of contemporary research. Accounts of the structures and reactions of organometallic complexes may be found in most inorganic chemistry texts (see Suggestions for Further Reading).

5–8 TRANSITION-METAL COMPLEXES AND LIVING SYSTEMS

Among the important transition-metal complexes in living systems are those that contain the porphyrin heterocyclic ring (Figure 5–25). Although porphyrin itself does not exist in nature, derivatives of it include the important natural products hemoglobin, chlorophyll, and cytochromes. Petroleum also contains porphyrins, which suggests that petroleum was derived from primordial living organisms.

The porphyrins are flat molecules that can act as tetradentate chelating groups (bonding through N atoms) with ions such as Mg^{2+}, Fe^{2+}, Fe^{3+}, Zn^{2+}, and Ni^{2+} in square planar complexes, as shown in Figure 5–26. The iron (as Fe^{2+} or Fe^{3+}) complex with the organic side chains shown in Figure 5–27 is called *heme*. The Mg^{2+} complex of porphyrin, with the side chains shown in Figure 5–28, is *chlorophyll*.

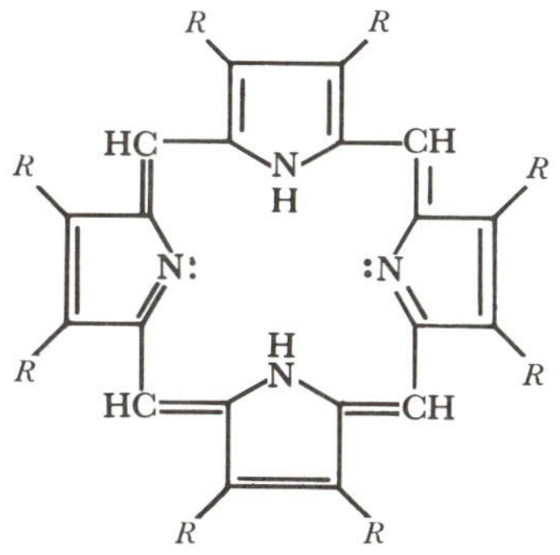

5–25
The porphyrin ring. Different porphyrins have different *R* groups bonded to the eight outermost positions of the ring.

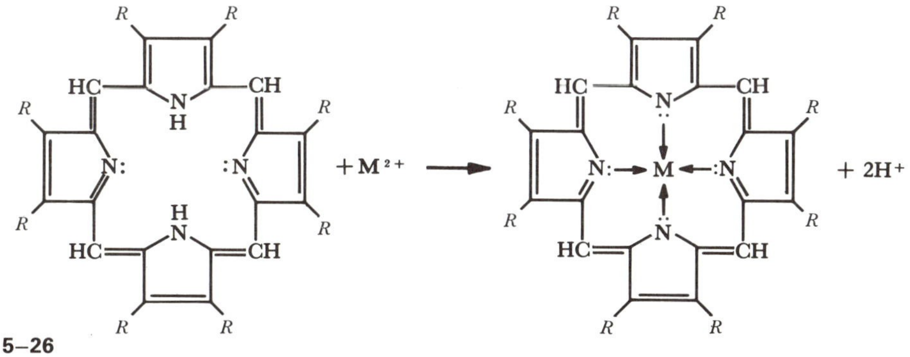

5–26
A porphyrin molecule can act as a tetradentate chelating group for metal ions such as Mg^{2+}, Fe^{2+}, Zn^{2+}, and Cu^{2+}.

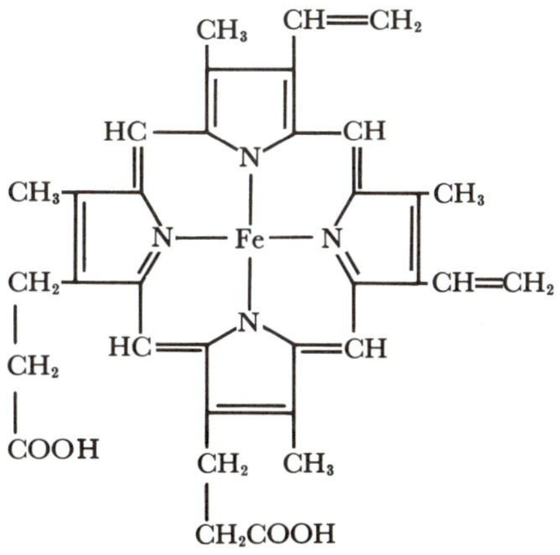

5–27
The iron porphyrin complex with the side chains as shown here is called a heme group.

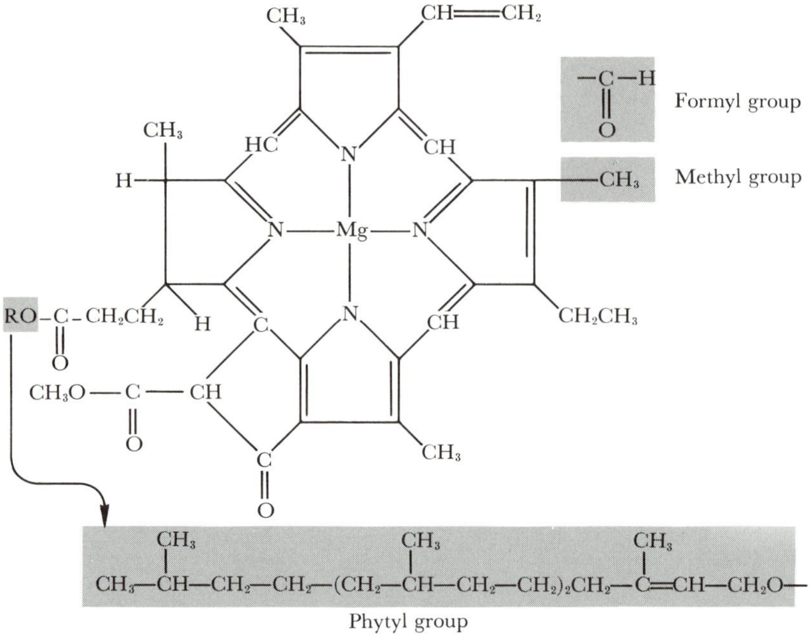

Formyl group

Methyl group

Phytyl group

5–28
The Mg^{2+} porphyrin derivative is called chlorophyll, and is the essential molecule in photosynthesis. Chlorophyll *a* is shown here; chlorophyll *b* has a formyl group in place of the methyl group.

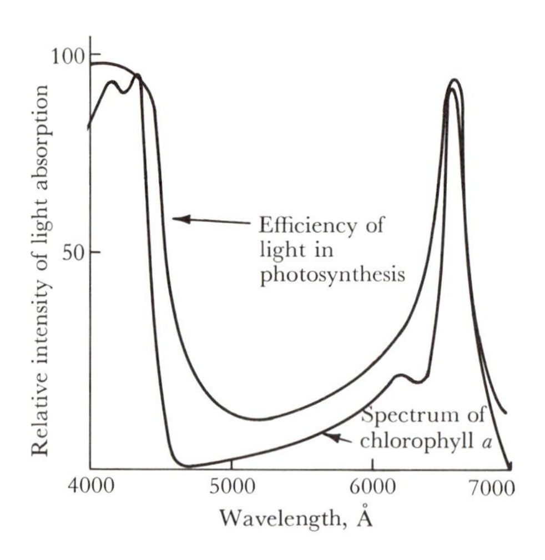

5–29
Chlorophyll *a* absorbs visible light except in the region around 5000 Å (green light), and thus appears green.

These two compounds, heme and chlorophyll, are the key components in the elaborate mechanism by which solar energy is trapped and converted for use by living organisms. We already have explained visible-light absorption of transition-metal complexes in terms of their closely spaced *d* levels. The porphyrin ring around the Mg^{2+} ion in chlorophyll also has closely spaced electronic energy levels, because the molecule contains an extensive network of conjugated double bonds. Chlorophyll molecules in plants can be electronically excited by absorbing photons of visible light (Figure 5–29). Thus chlorophyll is able to trap light and to use its energy to initiate a chain of chemical syntheses that ultimately produces sugars from carbon dioxide and water:

$$6CO_2 + 6H_2O \xrightarrow{h\nu} \underset{\text{glucose}}{C_6H_{12}O_6} + 6O_2$$

Scientists now believe that life evolved on earth in the presence of a reducing atmosphere, an atmosphere with ammonia, methane, water, and carbon dioxide, but no free oxygen. Free oxygen would degrade organic compounds faster than they could by synthesized by natural processes (electrical discharge, ultraviolet radiation, heat, or natural radioactivity). In the absence of free oxygen, organic compounds would accumulate in the oceans for eons until finally a localized bit of a chemical accumulated, which we would call "living."

Once developed, living organisms would exist by degrading naturally occurring organic compounds for their energy. The amount of life on earth would be limited severely if this degradation were the only source of energy. However, around three billion years ago the right combination of metal and porphyrin occurred and an entirely new source of energy was tapped—the sun. The first step that lifted life on earth above the role of a scavenger of high-energy organic compounds was an application of coordination chemistry.

Unfortunately, *photosynthesis* (as the chlorophyll photon-trapping process is called) liberates a dangerous by-product, oxygen. Oxygen was not only useless to these early organisms, it competed with them by oxidizing the naturally occurring organic compounds before they could be oxidized within the metabolism of the organisms. Oxygen was a far more efficient scavenger of high-energy compounds than living matter was. Even worse, the ozone (O_3) screen that slowly developed in the upper atmosphere cut off the supply of ultraviolet radiation from the sun and made the natural synthesis of more organic compounds even slower. From all contemporary points of view the appearance of free oxygen in the atmosphere was a disaster.

As so often happens, life bypassed the obstacle and turned a disaster into an advantage. The waste products of the original simple organisms had been compounds such as lactic acid and ethanol, which can release large amounts of energy if oxidized completely to CO_2 and H_2O. Living organisms evolved that were able to convert the poisonous O_2 to H_2O and CO_2, and to gain the energy of combustion of what were once its waste products. Thus aerobic metabolism had evolved.

Again, the significant development was an advance in coordination chemistry. The central components in the new terminal oxidation chain, by which the combustion of organic molecules was brought to completion, are the *cytochromes*. These are molecules in which an iron ion is complexed with a porphyrin to make a heme (Figure 5–27), and the heme is surrounded with protein. The iron changes from Fe^{2+} to Fe^{3+} and back again as electrons are transferred from one component in the chain to another. The entire terminal oxidation chain is a carefully interlocked set of oxidation-reduction reactions, in which the overall result is the reverse of the photosynthetic process:

$$6O_2 + C_6H_{12}O_6 \longrightarrow 6CO_2 + 6H_2O$$

The energy liberated is stored in the organism for use as needed. The entire chlorophyll-cytochrome system can be regarded as a mechanism for converting the energy of solar photons into stored chemical energy in the muscles of living creatures.

Iron atoms usually exhibit octahedral coordination. What occupies the two coordination positions above and below the plane of the porphyrin ring? In cytochrome *c*, the heme group sits in a crevice in the surface of the protein molecule (Figure 5–30). From each wall of this crevice one additional ligand extends toward the heme: on one side a nitrogen lone electron pair from a histidine side chain on the protein, and on the other side a sulfur lone pair from a methionine side chain (Figure 5–31). Therefore the bonds from the iron atom are directed octahedrally to five nitrogen atoms and one sulfur atom. The ligands around the iron in the complex, and the protein wrapped around the whole structure, allow the cytochrome *c* molecule to transfer electrons efficiently in the terminal oxidation chain.

There is one more step in the story of metal-porphyrin complexes. With the guarantee of new energy sources, multicelled organisms evolved. At this point arose the problem, not of obtaining food or oxygen, but of transporting oxygen to the proper place in the organism. Simple gaseous diffusion through body fluids will work for small organisms, but not for large, multicelled creatures. Again, a natural limit was placed on evolution.

5–30▶

Cytochrome *c* is a globular protein with 104 amino acids in one protein chain and an iron-containing heme group. In this schematic drawing, each amino acid is represented by a numbered sphere. The heme group is seen nearly edgewise in a vertical crevice in the molecule. Copyright © 1972 Richard E. Dickerson and Irving Geis, from *Scientific American*, p. 62, April 1972.

AMINO END
1 Gly
4 Glu
90 Glu
5 Lys
8 Lys
Gln 12
Lys 87
Lys 7
9 Ile
89 Thr
Asp 93
Asp 2
Val 3
11 Val
86 Lys
Ile 85
Gly 6
88 Lys
Lys 13
10 Phe
92 Glu
Cys 14
Leu 94
96 Ala
16 Gln
Gly 84
Arg 91
Tyr 97
Ala 83
Lys 100
Ile 95
68 Leu
Glu 69
HEME
98
Ala 15
65 Met
81 Ile
Ala 101
Phe 82
20 Val
Leu 64
Cys 17
Lys 99
Asn 70
21 Glu
Met 80
Fe
His 18
Thr 19
Lys 22
103 Asn
Glu 66
Lys 27
28 Thr
24 Gly
His 33
Leu 32
Thr 63
Glu 61
71 Pro
67 Tyr
Asn 31
72 Lys
25 Lys
Phe 36
35 Leu
Gly 29
78 Thr
60 Lys
23 Gly
Lys 73
Glu 62
75 Ile
Gly 34
CARBOXYL END
26 His
30 Pro
Tyr 74
79 Lys
104 Glu
59 Trp
37 Gly
Pro 76
77 Gly
41 Gly
58 Thr
38 Arg
46 Phe
52 Asn
57 Ile
40 Thr
42 Gln
Ala 51
Tyr 48
Thr 47
49 Thr
44 Pro
56 Gly
39 Lys
Ala 43
Lys 55
45 Gly
53 Lys
Asn 54
Asp 50

Methionine

Histidine

Heme group seen on edge

5–31
The iron atom in cytochrome *c* is octahedrally coordinated through five bonds to nitrogen atoms and one to a sulfur atom. One nitrogen atom and the sulfur atom come from side groups on the protein chain. The other four nitrogen atoms are on the porphyrin ring of the heme.

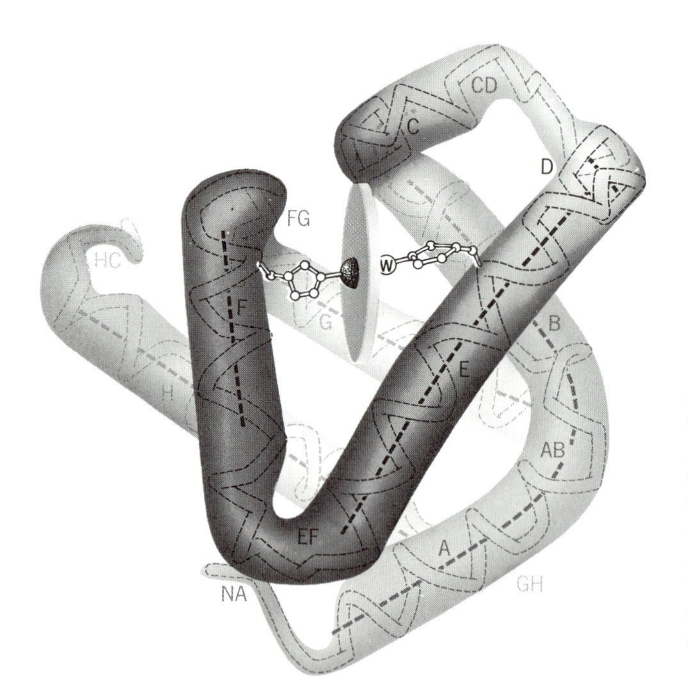

5–32
The myoglobin molecule is a storage unit for an oxygen molecule in muscle tissue. The heme group is represented by a flat disk. The iron atom is a ball at the center. The circled W marks the bonding site for O_2. The path of the polypeptide chain is shown by double dashed lines. Figures 5–32 and 5–33 copyright © 1969 by Richard E. Dickerson and Irving Geis.

5–33
The hemoglobin molecule is the carrier of oxygen in the bloodstream. It is built from four subunits, each of which is constructed like a myoglobin molecule. This figure and that of myoglobin are reprinted from Richard E. Dickerson and Irving Geis, *The Structure and Action of Proteins*, Harper and Row, New York, 1969.

Once again, the way out of the impasse was found with coordination chemistry. Molecules of iron, porphyrin, and protein evolved, in which Fe^{2+} could bind a molecule of oxygen without being oxidized to Fe^{3+} by it. The oxidation to Fe^{3+} was, in a sense, "aborted" after the first binding step. Oxygen merely was carried along to be released under the proper conditions of acidity and oxygen scarcity. Two compounds evolved, *hemoglobin*, which carries oxygen in the blood, and *myoglobin*, which receives and stores oxygen in the muscles until it is needed in the cytochrome process.

The myoglobin molecule is depicted in Figure 5–32. As in cytochrome *c*, four of the six octahedral iron positions are taken by heme nitrogens. The nitrogen of a histidine is bonded in the fifth position. However, the sixth position has no ligand. This is the place where the oxygen molecule bonds, marked by the circled W. In myoglobin, the iron is in the Fe^{2+} state. If the iron is oxidized to Fe^{3+}, the molecule is inactivated and a water molecule occupies the oxygen position.

Hemoglobin is a package of four myoglobinlike molecules (Figure 5–33). From x-ray crystallographic studies it has become apparent that the four subunits of hemoglobin shift by 7 Å relative to one another when oxygen bonds. Hemoglobin and myoglobin now become a model system for transition-metal chemists to study. Why does bonding at the sixth ligand site of the iron complex cause the protein subunits to rearrange? Why does the oxygen molecule fall away from hemoglobin in an acid environment (such as in oxygen-poor muscle tissue)? How is the coordination chemistry of hemoglobin and myoglobin so carefully meshed that myoglobin binds oxygen just as hemoglobin releases it at the tissues? These are questions that probably will be answered from further research.

Heme also is at the active sites in enzymes that decompose H_2O_2 to H_2O and O_2. Manganese, cobalt, copper, and molybdenum, among others, also are essential transition metals in enzyme catalysis. With the evolution of myoglobin and hemoglobin, the size limitation was removed from living organisms. Thereafter, all of the multicelled animals that we ordinarily see around us evolved. In the sense that transition metals and double-bonded organic ring systems such as porphyrin are uniquely suited for absorbing visible light, and their combinations have a particularly rich redox chemistry, life indeed depends on coordination chemistry.

SUGGESTIONS FOR FURTHER READING

F. Basolo and R. Johnson, *Coordination Chemistry*, Benjamin, Menlo Park, Calif., 1964.

F. A. Cotton and G. Wilkinson, *Advanced Inorganic Chemistry*, 2nd. ed., Wiley, New York, 1966.

R. E. Dickerson, "The Structure and History of an Ancient Protein," *Scientific American* (April, 1972).

H. B. Gray, *Electrons and Chemical Bonding*, Benjamin, Menlo Park, Calif., 1965.

L. E. Orgel, *An Introduction to Transition Metal Chemistry*, Wiley, New York, 1966.

E. G. Rochow, *Organometallic Chemistry*, Reinhold, New York, 1964.

QUESTIONS AND PROBLEMS

1. Why are octahedral complexes with d^3 and d^6 configurations particularly stable? Which electronic configurations would you predict to be more important for stability in high-spin complexes? In low-spin complexes?
2. All octahedral complexes of V^{3+} have the same number of unpaired electrons, no matter what the nature of the ligand. Why is this so?
3. How does the ligand field theory account for the order of ligands in the spectrochemical series?

4. What is a chelate? If porphyrin is a tetradentate chelating group, and ethylenediamine is a bidentate chelating group, how would triethylenetetraamine, diethylenetriamine, and ethylenediaminetetraacetate (EDTA) be described?
5. Explain why $Co(CN)_6^{3-}$ is extremely stable but $Co(CN)_6^{4-}$ is not.
6. Explain the fact that most complexes of Zn^{2+} are colorless.
7. Explain why octahedrally coordinated Mn^{3+} is very unstable, whereas octahedral complexes of Cr^{3+} are extremely stable.
8. Using ligand field theory predict the number of unpaired electrons in the following complexes: FeO_4^{2-}, $Mn(CN)_6^{3-}$, $NiCl_4^{2-}$ (tetrahedral), $PdCl_4^{2-}$ (square planar), $MnCl_4^{2-}$, $Co(en)_3^{2+}$, $Co(en)_3^{3+}$, $Rh(NH_3)_6^{3+}$, $CoBr_4^{2-}$, and $Pt(NH_3)_4^{2+}$.
9. Carbon monoxide is a strong-field ligand that stabilizes transition metals in unusually low oxidation states. For example, $V(CO)_6$ and $V(CO)_6^-$ both are stable complexes. What are the ground-state electronic configurations of these two complexes in the ligand-field levels t_{2g} and e_g? Which member of the series $V(CO)_6$, $Cr(CO)_6$, and $Mn(CO)_6$ would you expect to be most stable? Which would be least stable? Why?
10. One of the most toxic substances known to man is tetracarbonylnickel(0), $Ni(CO)_4$. Predict its geometrical structure, using localized molecular orbital theory. Formulate the ground-state electronic structure of $Ni(CO)_4$ using ligand field theory. Would you expect to observe *d–d* transitions in this compound? Why or why not?
11. A complex widely used in studying reactions of octahedral complexes is $Co(NH_3)_5Cl^{2+}$. Actually, this complex is octahedral only in an approximate sense, because NH_3 and Cl^- have different ligand-field strengths. How would you modify the ligand-field energy levels of $Co(NH_3)_6^{3+}$ in formulating a splitting diagram for $Co(NH_3)_5Cl^{2+}$? (For convenience, place the Cl^- ligand along the Z axis of a Cartesian coordinate system.) Apply your theory to explain the fact that $Co(NH_3)_6^{3+}$ is yellow (absorbs light at 4300 Å), whereas $Co(NH_3)_5Cl^{2+}$ is purple (absorbs light at 5300 Å). Which d orbital receives the electron in the excitation giving rise to the 5300-Å band in $Co(NH_3)_5Cl^{2+}$? Why?
12. If $Mn(H_2O)_6^{2+}$, $Fe(H_2O)_6^{2+}$, and $Co(H_2O)_6^{2+}$ had been low-spin, how would Figure 5–8 have appeared?
13. Why are Ni^{2+} complexes with weak-field ligands octahedral and those with strong-field ligands square planar?
14. Electronically excited molecules often emit light just as do excited atoms. However, excited molecules also may use their excess energy to break chemical bonds and thereby undergo chemical reactions that might not occur otherwise. An example is the metal complex $W(CO)_6$. The molecule is unreactive in its ground state, but upon irradiation by light of 3000-Å wavelength the following reaction occurs:

$$W(CO)_6 \rightarrow W(CO)_5 + CO$$

From this information estimate an upper limit, in kcal mole^{-1}, for the W–CO bond energy. As the reaction proceeds the concentration of carbon monoxide increases. Is there a possibility that the secondary reaction

$$CO \xrightarrow{\lambda = 3000\text{ Å}} C + O$$

can occur? Why or why not?

15. a) Give the ligand-field electronic configuration of the ground state of $W(CO)_6$. Is the complex diamagnetic or paramagnetic? The first electronic absorption band occurs at about 31,000 cm^{-1}. Assign this band to an electric transition in the complex. Is the photochemical dissociation of $W(CO)_6$ to $W(CO)_5 + CO$, described in Problem 14, reasonable? Explain.

    b) The $W(CO)_5$ molecule has a square pyramidal structure. Assume a reference coordinate system in which only one CO ligand is along the Z axis and predict the ligand-field splitting for this complex. $W(CO)_5$ absorbs light strongly at 25,000 cm^{-1}. Assign the band to an electronic transition in $W(CO)_5$ and explain why the absorption is at lower energy than in $W(CO)_6$.

16. The $CuCl_2$ molecule has been observed in the gas phase. It has a linear structure. Assume that the internuclear line is along the Z axis and predict the ligand-field splitting diagram for $CuCl_2$. What is the ground-state electronic configuration of $CuCl_2$? How many d–d transitions should be observed? What are the transition assignments?

17. Consider the following complex ions: MnO_4^{3-}, $Pd(CN)_4^{2-}$, NiI_4^{2-}, $Ru(NH_3)_6^{3+}$, $MoCl_6^{3-}$, $IrCl_6^{2-}$, $AuCl_4^-$, and FeF_6^{3-}. Use ligand field theory to predict the structure and number of unpaired electrons in each ion.

18. Give reasonable examples of the following: (a) d^5 high-spin octahedral complex; (b) low-spin square planar complex; (c) d^0 tetrahedral complex; (d) d^5 low-spin octahedral complex; (e) d^3 complex.

19. (a) Calculate the energy in cm^{-1} of the absorption spectral line in isolated Li^{2+} corresponding to the transition $1s \rightarrow 3d$. Do the $1s \rightarrow 3s$ and $1s \rightarrow 3p$ transitions have the same energy? (b) Now assume that the Li^{2+} is placed in an octahedral ligand field. How many different electronic transitions would you expect to observe from the $1s$ orbital to the $n = 3$ orbitals in this ligand field? How many would be expected in a tetrahedral field? How many in a square planar field?

20. Which complex in each of the following pairs would you expect to be more stable: (a) $PtCl_4^{2-}$ or PtI_4^{2-}; (b) $Fe(H_2O)_6^{3+}$ or $Fe(NH_3)_6^{3+}$; (c) $FeCl_4^-$ or FeI_4^-; (d) $ZnCl_4^{2-}$ or $ZnBr_4^{2-}$; (e) $HgCl_4^{2-}$ or $HgBr_4^{2-}$; (f) $Ag(H_2O)_6^+$ or $Ag(NH_3)_2^+$; (g) BF_4^- or BCl_4^-; (h) $Cu(NH_3)_4^{2+}$ or $Cu(H_2O)_4^{2+}$.

21. Cobaltocene, $Co(C_5H_5)_2$, and nickelocene, $Ni(C_5H_5)_2$, have sandwich structures similar to that of ferrocene. Formulate the ground-state electronic structures for these two organometallic π complexes. How many unpaired electrons are there in each case? Would you expect cobaltocene to be more, or less, stable than ferrocene? Why?

22. Cobaltocene can be oxidized to the cobalticenium ion, $Co(C_5H_5)_2^+$. Would you expect this ion to be diamagnetic, or paramagnetic? Explain.

23. Predict which complex in each of the following pairs will have the lower-energy d–d transition: (a) $Co(NH_3)_5F^{2+}$ or $Co(NH_3)_5I^{2+}$; (b) $Co(NH_3)_5Cl^{2+}$ or $Co(NH_3)_5NO_2^{2+}$; (c) $Pt(NH_3)_4^{2+}$ or $Pd(NH_3)_4^{2+}$; (d) $Co(CN)_6^{3-}$ or $Ir(CN)_6^{3-}$; (e) $Co(CN)_5H_2O^{2-}$ or $Co(CN)_5I^{3-}$; (f) $V(H_2O)_6^{2+}$ or $Cr(H_2O)_6^{3+}$; (g) $RhCl_6^{3-}$ or $Rh(CN)_6^{3-}$; (h) $Ni(H_2O)_6^{2+}$ or $Ni(NH_3)_6^{2+}$.

24. Predict whether a reaction will occur when Br^- is added to aqueous solutions of each of the following complex ions: FeF_6^{3-}, $PdCl_4^{2-}$, HgI_4^{2-}, BF_4^-, $Pt(NH_3)_3I^+$, $Co(NH_3)_5I^{2+}$, $Co(NH_3)_5F^{2+}$, and $Pt(NH_3)_3H_2O^{2+}$.

25. Many complexes of Cu^{2+} have square planar structures. What is the ground-state electronic configuration of a square planar Cu^{2+} complex? How many *d–d* transitions of different energies can be expected?

26. A common inorganic chemical compound, ferric ammonium sulfate, has the formula $Fe_2(SO_4)_3(NH_4)_2SO_4{\cdot}24H_2O$. Large crystals of the compound are very pale violet, due to weak absorptions (ε values between 0.05 and 1) in the visible region of the spectrum. The Fe^{3+} in the compound is present as the hexaaquo complex ion, $Fe(H_2O)_6^{3+}$. Using ligand field theory formulate an explanation of the weak absorption bands.

27. The ferrocyanide ion, $Fe(CN)_6^{4-}$, does not exhibit an absorption band in the visible region, but ferricyanide, $Fe(CN)_6^{3-}$, absorbs strongly at approximately 25,000 cm^{-1}. What type of electronic transition is responsible for this strong absorption, which makes $Fe(CN)_6^{3-}$ red?

28. Explain the fact that the lowest $L \rightarrow M$ charge-transfer band shifts from 18,000 cm^{-1} in MnO_4^- to 26,000 cm^{-1} in CrO_4^{2-}.

29. Would you expect a given $M \rightarrow L$ charge-transfer transition to be at lower, or higher, energy in $Cr(CO)_6$ than in $Mn(CO)_6^+$? Why?

30. The yellow complex HgI_4^{2-} exhibits a strong absorption band at $\bar{\nu}_{max} = 27{,}400\ cm^{-1}$. What type of electronic transition is responsible for this band?

6

Bonding in Solids

In previous chapters we explored in some detail the bonding between atoms in isolated molecules. In many solids discrete molecules exist, but in many others atoms are bonded together in an infinite array to build a "giant" molecule. In this chapter we will relate the properties of solid substances to the types of bonding interactions between the atoms or molecules present.

6–1 TYPES OF SOLIDS

Solids that are built by weak attractive interactions between individual molecules are called *molecular solids*. Examples of molecular solids include iodine crystals, composed of discrete I_2 molecules, and paraffin wax, composed of long-chain alkane molecules. At very low temperatures the noble gases exist as molecular solids that are held together by weak interatomic forces. For example, argon freezes at $-189°C$ to make the close-packed structure shown in Figure 6–1. Examples of nonpolar molecules that crystallize at low temperatures to give molecular solids include Br_2, which freezes at $-7°C$ to build the structure shown in Figurc 6–2. Methane, CH_4, freezes at $-183°C$ to form the close-packed crystal shown in Figure 6–3.

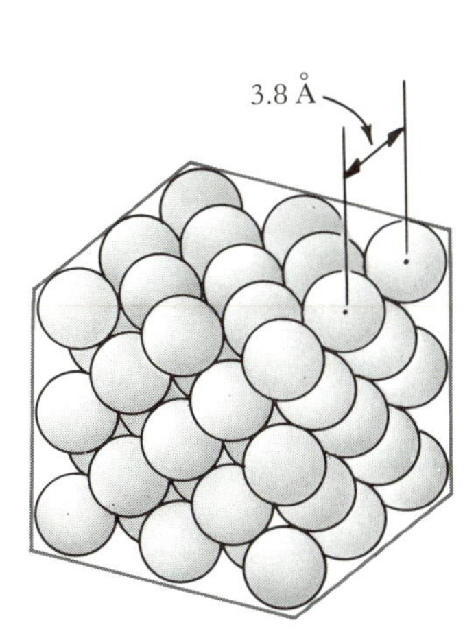

6–1
The structure of solid argon. Each sphere represents an individual Ar atom, in cubic close packing with 3.8 Å between atomic centers.

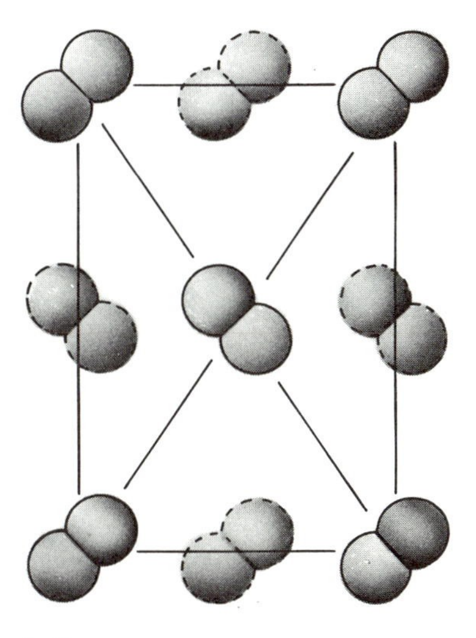

6–2
The structure of crystalline bromine, Br_2. The solid outlines indicate one layer of packed molecules, and the dashed outlines indicate a layer beneath. The molecules have been shrunk for clarity in this drawing; they are actually in close contact within a layer, and the layers are packed against one another.

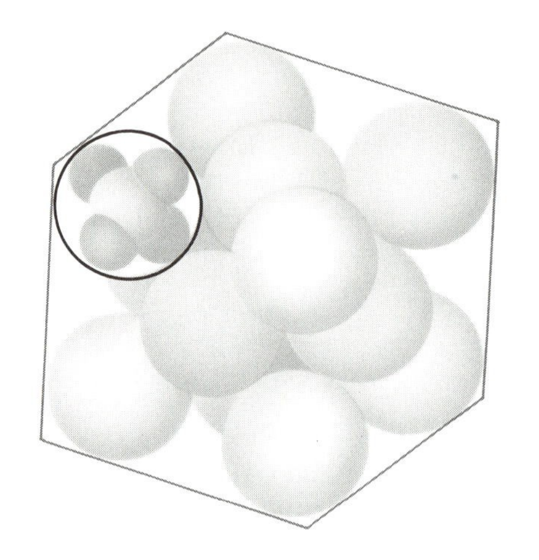

6–3
The structure of solid methane, CH_4. Each large sphere represents a methane molecule, as indicated at the upper left. The methane molecules are arranged in cubic close packing.

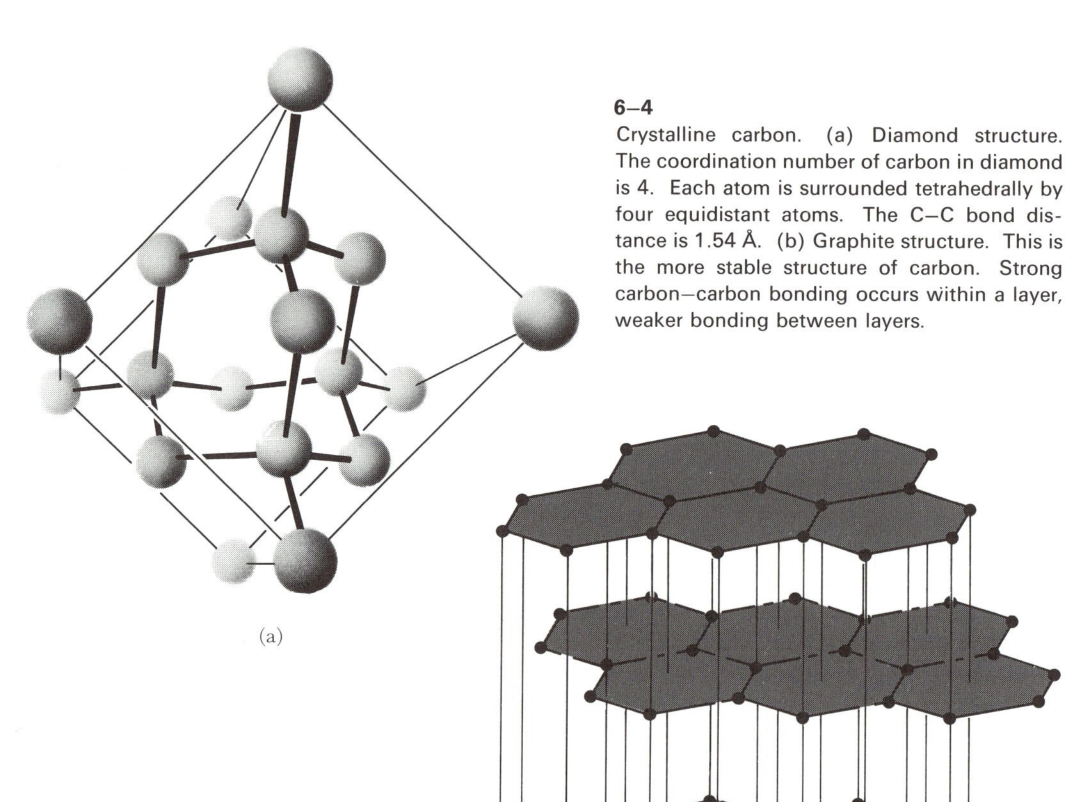

6–4
Crystalline carbon. (a) Diamond structure. The coordination number of carbon in diamond is 4. Each atom is surrounded tetrahedrally by four equidistant atoms. The C–C bond distance is 1.54 Å. (b) Graphite structure. This is the more stable structure of carbon. Strong carbon–carbon bonding occurs within a layer, weaker bonding between layers.

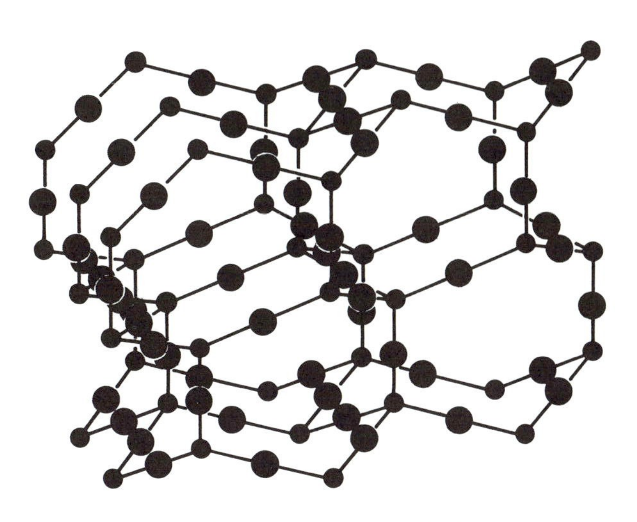

6–5
The three-dimensional network of silicate tetrahedra in one crystalline form of silicon dioxide, $(SiO_2)_n$.

Nonmetallic network solids consist of infinite arrays of bonded atoms; no discrete molecules can be distinguished. Thus any given piece of a network solid may be considered a giant, covalently bonded molecule. Network solids generally are poor conductors of heat and electricity. Strong covalent bonds among neighboring atoms throughout the structure give these solids strength and high melting temperatures. Some of the hardest substances known are nonmetallic network solids.

Diamond, the hardest allotrope of carbon, has the network structure shown in Figure 6–4(a). Diamond sublimes (volatilizes directly to a gas), rather than melts, at 3500°C and above. Graphite, a softer allotrope of carbon, has the layered structure shown in Figure 6–4(b). Silicon dioxide is another high-melting (>1600°C) three-dimensional network solid (Figure 6–5).

One feature that distinguishes network solids from metals is the lower coordination number of atoms in network structures. In the preceding examples the coordination number of C, in diamond, and Si is four, and that of O in $(SiO_2)_n$. is two. In Section 6–3, we will see that a localized molecular orbital picture of bonding satisfactorily accounts for the properties of diamond.

Metallic solids also consist of infinite arrays of bonded atoms, but in contrast to nonmetals each atom in a metal has a high coordination number: sometimes four or six, but more often eight or twelve (Figure 6–21). The band theory of delocalized molecular orbitals will be developed in Section 6–3 to explain the fact that metals generally are good conductors of electricity.

In the periodic table shown in Figure 6–6 the elemental solids are classified as metallic, network nonmetallic, or molecular. The majority of elements

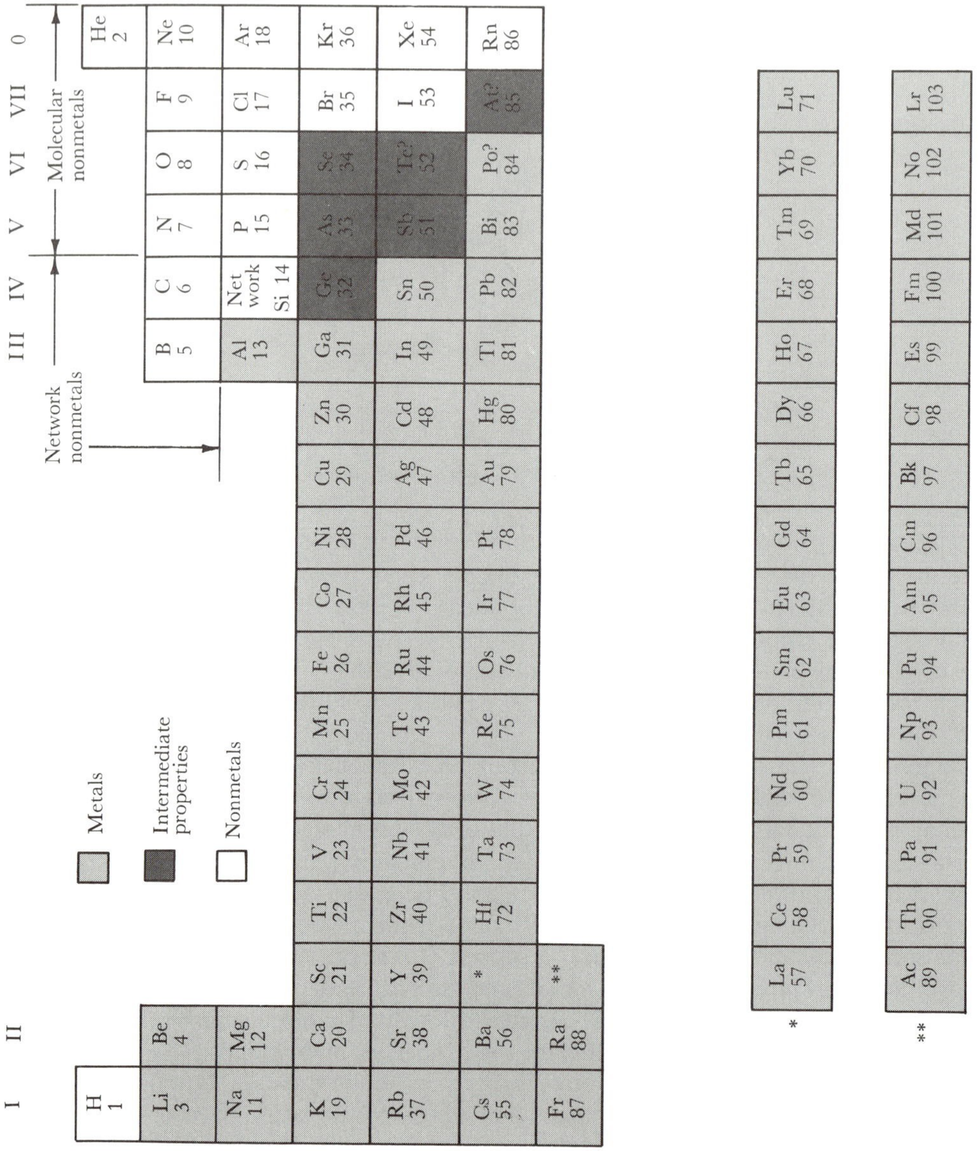

Metals | Intermediate properties | Nonmetals

Network nonmetals | Molecular nonmetals

I	II											III	IV	V	VI	VII	0
H 1																	He 2
Li 3	Be 4											B 5	C 6	N 7	O 8	F 9	Ne 10
Na 11	Mg 12											Al 13	Net work Si 14	P 15	S 16	Cl 17	Ar 18
K 19	Ca 20	Sc 21	Ti 22	V 23	Cr 24	Mn 25	Fe 26	Co 27	Ni 28	Cu 29	Zn 30	Ga 31	Ge 32	As 33	Se 34	Br 35	Kr 36
Rb 37	Sr 38	Y 39	Zr 40	Nb 41	Mo 42	Tc 43	Ru 44	Rh 45	Pd 46	Ag 47	Cd 48	In 49	Sn 50	Sb 51	Te? 52	I 53	Xe 54
Cs 55	Ba 56	*	Hf 72	Ta 73	W 74	Re 75	Os 76	Ir 77	Pt 78	Au 79	Hg 80	Tl 81	Pb 82	Bi 83	Po? 84	At? 85	Rn 86
Fr 87	Ra 88	**															

*	La 57	Ce 58	Pr 59	Nd 60	Pm 61	Sm 62	Eu 63	Gd 64	Tb 65	Dy 66	Ho 67	Er 68	Tm 69	Yb 70	Lu 71
**	Ac 89	Th 90	Pa 91	U 92	Np 93	Pu 94	Am 95	Cm 96	Bk 97	Cf 98	Es 99	Fm 100	Md 101	No 102	Lr 103

6–6
Structural characteristics of elemental solids.

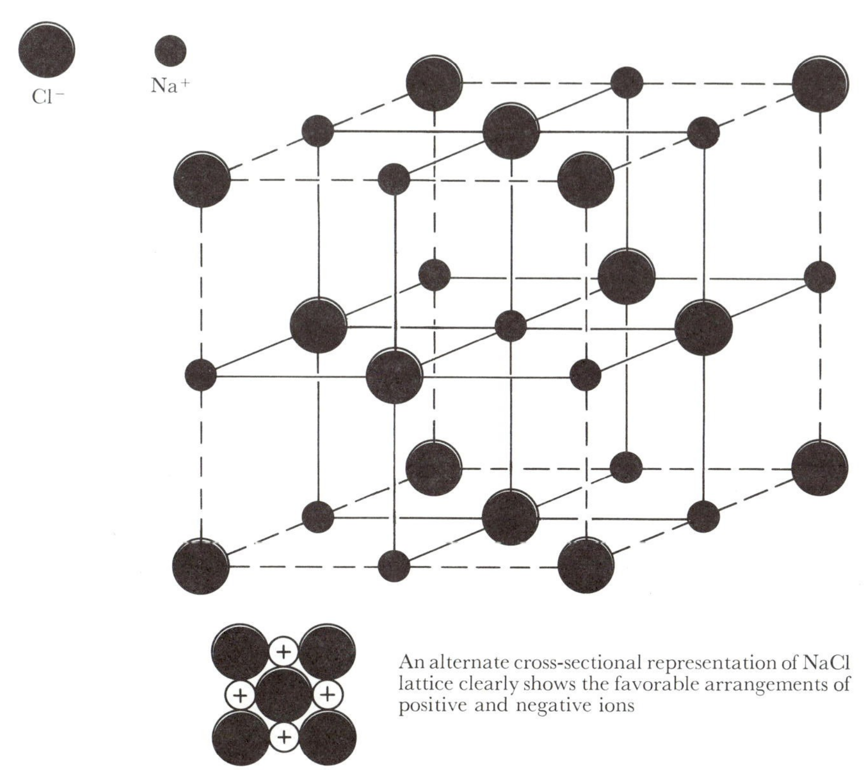

6–7
Representation of the ionic NaCl structure. The bottom figure is a representation of a cross section of the NaCl structure.

crystallize in metallic structures in which each atom has a high coordination number. Included as metals are elements such as tin and bismuth, which crystallize in structures with relatively low atomic coordination numbers but which still have strong metallic properties. The gray area of the periodic table includes elements that have borderline properties. Although germanium crystallizes in a diamondlike structure in which the coordination number of each Ge atom is only four, certain of its properties resemble those of metals. This similarity to metals indicates that the valence electrons in germanium are not held as tightly as would be expected in a true nonmetallic network solid. Arsenic, antimony, and selenium exist as either molecular or metallic solids, although the so-called metallic structures have relatively low atomic coordination numbers. We know that tellurium crystallizes in a metallic structure, and it seems reasonable to predict that it also may exist as a molecular solid. From its position in the periodic table we predict intermediate properties for astatine, which has not been studied in detail.

Ionic solids consist of infinite arrays of positive and negative ions that are held together by electrostatic forces. These forces are the same as those that hold a molecule of NaCl together in the vapor phase (Section 2–8). In solid NaCl the Na^+ and Cl^- ions are arranged to maximize the electrostatic attraction, as shown in Figure 6–7. The coordination number of each Na^+ ion is six, and each Cl^- ion similarly is surrounded by six Na^+ ions. Because ionic bonds are very strong, much energy is required to break down the structure in solid-to-liquid or liquid-to-gas transitions. Thus ionic compounds have high melting and boiling temperatures.

The preceding discussion has distinguished four types of solids—molecular, network nonmetallic, metallic, and ionic. Of these types, by far the weakest bonding is found in molecular solids, in which only *intermolecular* forces hold the crystal together. In the next section we will examine in more detail the nature of these intermolecular forces.

6–2 MOLECULAR SOLIDS

Molecules such as H_2, N_2, O_2, and F_2 form molecular solids because all the valence orbitals are used either for *intramolecular* bonding or are occupied with nonbonding electrons. Thus any intermolecular bonding that holds molecules together in the solid must be weak compared with the strength of the intramolecular bonding in the molecules. The weak forces that contribute to intermolecular bonding are called van der Waals forces.

Van der Waals forces

There are two principal van der Waals forces. The most important force at short range is the repulsion between electrons in the filled orbitals of atoms on neighboring molecules. This electron-pair repulsion is illustrated in Figure 6–8. The analytical expression commonly used to describe the energy resulting from this interaction is

$$\text{van der Waals repulsion energy} = be^{-aR} \qquad (6\text{–}1)$$

in which b and a are constants for two interacting atoms. Notice that this repulsion term is very small at large values of the interatomic distance, R.

The second force is the attraction that results when electrons in the occupied orbitals of the interacting atoms synchronize their motion to avoid each other as much as possible. For example, as shown in Figure 6–9, electrons in orbitals of atoms belonging to interacting molecules can synchronize their motion to produce an instantaneous dipole-induced dipole attraction. If at any instant the left atom in Figure 6–9 had more of its electron density at the left, as shown, then the atom would be a tiny dipole with a negative left side and a positive right side. This positive side would attract electrons on the right atom in the figure and would change this atom into a dipole with similar orientation.

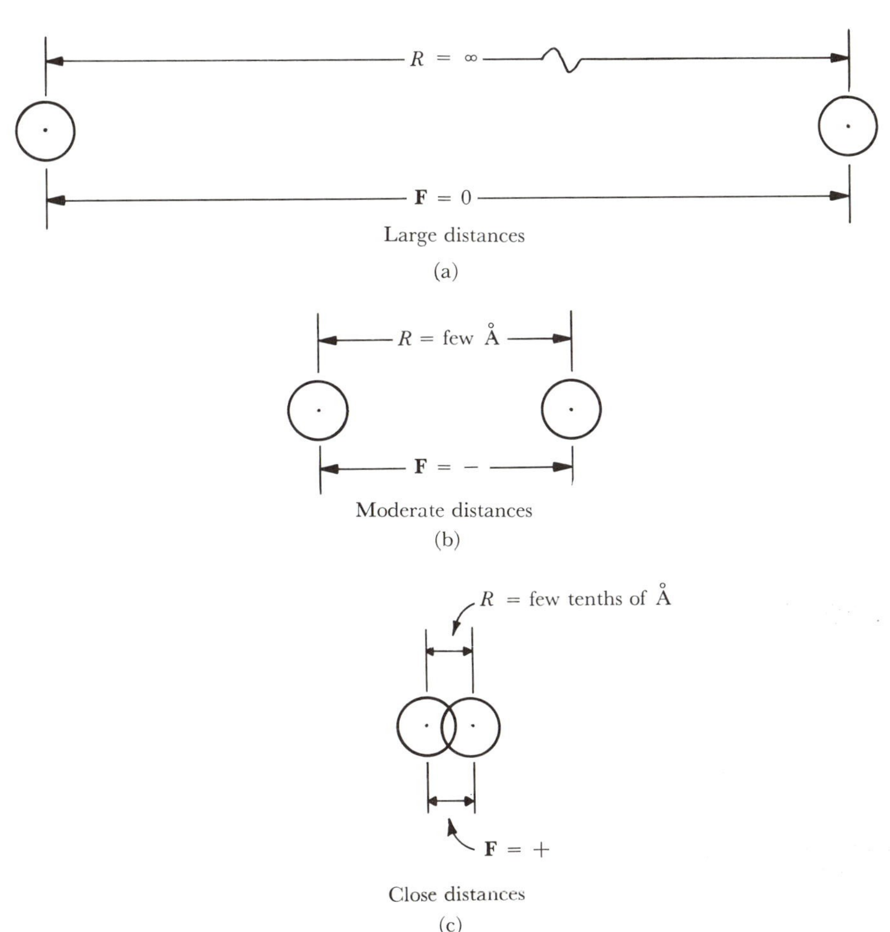

6–8
Repulsion of electrons in filled orbitals. (a) At very large distances two atoms or molecules behave toward each other as neutral species and neither repel nor attract one another. The force between them, **F**, is zero. (b) At moderate distances two atoms or molecules have not yet come close enough for repulsion to be appreciable. However, they do attract one another (see Figure 6–9) because of deformations of their electron densities. (c) At close range, when the electron density around one atom or molecule is large in the same region of space as the electron density around the other atom or molecule (i.e., when the filled orbitals overlap), coulomblike repulsion dominates and the two molecules repel one another.

Therefore these two atoms would attract each other because the positive end of the left atom and the negative end of the right atom are close. Similarly, fluctuation in electron density of the right atom will induce a temporary dipole, or asymmetry of electron density, in the left atom. The electron densities are fluctuating continually, yet the net effect is an extremely small but important attraction between atoms. The energy resulting from this attractive force is

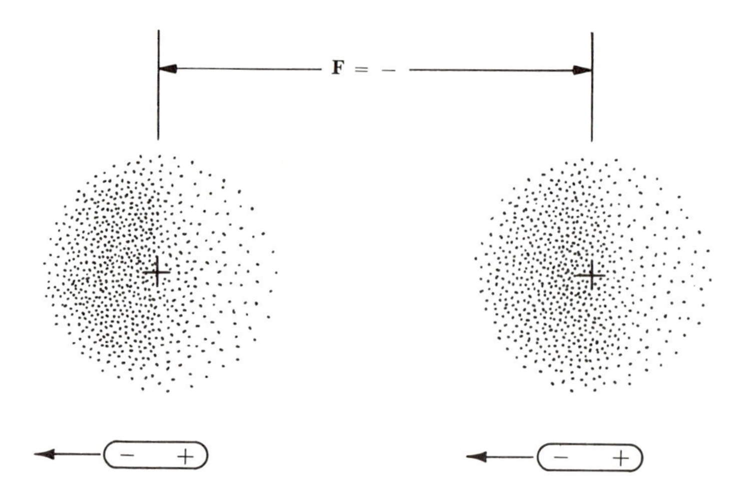

Instantaneous polarization of an atom leaves, for a moment, more electron density on the left than on the right, thus creating an "instantaneous dipole"

This "instantaneous dipole" can polarize another atom by attracting more electron density to the left, thereby creating an "induced dipole"

6–9
Schematic illustration of the instantaneous dipole-induced dipole interaction that gives rise to a weak attraction. For the brief instant that this figure describes, there is an attractive force, **F**, between the instantaneous dipole and the induced dipole. The effect is reciprocal; each atom induces a polarization in the other.

known as the *London energy*, after Fritz London, who derived the quantum mechanical theory for this attraction in 1930. The London energy varies inversely with the sixth power of the separation between atoms:

$$\text{London energy} = -\frac{d}{R^6} \qquad (6\text{–}2)$$

in which d is a constant and R is the distance between atoms. This "inverse sixth" attractive energy decreases rapidly with increasing R, but not nearly as rapidly as the van der Waals repulsion energy. Thus at longer distances the London attraction is more important than the van der Waals repulsion, consequently a small net attraction results.

The total potential energy of van der Waals interactions is the sum of the attractive energy of Equation 6–2 and the repulsive energy of Equation 6–1:

$$PE = be^{-aR} - \frac{d}{R^6} \qquad (6\text{–}3)$$

Table 6–1. Van der Waals energy parameters

Interaction pair	a (au)$^{-1}$ [a]	b, kcal mole^{-1}	d, kcal mole^{-1} (au)^6
He—He	2.10	4.1×10^3	1.5×10^3
He—Ne	2.27	20.7×10^3	2.9×10^3
He—Ar	2.01	30.0×10^3	9.7×10^3
He—Kr	1.85	16.4×10^3	13.7×10^3
He—Xe	1.83	26.6×10^3	21.3×10^3
Ne—Ne	2.44	104.8×10^3	5.7×10^3
Ne—Ar	2.18	151.8×10^3	19.2×10^3
Ne—Kr	2.02	82.8×10^3	26.7×10^3
Ne—Xe	2.00	134.2×10^3	41.5×10^3
Ar—Ar	1.95	219.5×10^3	64.6×10^3
Ar—Kr	1.76	119.8×10^3	90.1×10^3
Ar—Xe	1.74	194.4×10^3	139.3×10^3
Kr—Kr	1.61	65.2×10^3	125.4×10^3
Kr—Xe	1.58	106.0×10^3	194.4×10^3
Xe—Xe	1.55	171.8×10^3	301.1×10^3

[a] 1 au = 1 atomic unit = 0.529 Å. The value of R in Equation 6–3 must be expressed in atomic units as well.

The total van der Waals potential energy can be compared quantitatively with ordinary covalent bond energies by examining systems for which the curves of potential energy versus interatomic distance, R, are known accurately. We can calculate values for the constants a, b, and d from experimental data on the deviation of real gases from ideal gas behavior. Some of these values for interactions of noble gases are listed in Table 6–1.

The potential energy curve for van der Waals interactions between helium atoms is illustrated in Figure 6–10. At separations of more than 3.5 Å, the second term in Equation 6–3 predominates. As the atoms move closer together they attract each other more, and the energy of the system decreases. However, at distances closer than 3 Å the strong electron-pair repulsion overwhelms the London attraction, and the potential energy curve in Figure 6–10 rises. A balance between attraction and repulsion exists at 3-Å separation, and the He---He "molecule" is 18.2 cal mole^{-1} more stable than two isolated atoms.

Figure 6–10 also shows the marked contrast between van der Waals attraction and covalent bonding. In the H_2 molecule strong electron–proton attractions in the bonding molecular orbital cause the potential energy to decrease as the H atoms approach one another, and it is proton–proton repulsion that makes the energy increase sharply if the atoms are pushed too closely together. This proton–proton repulsion operates at smaller distances

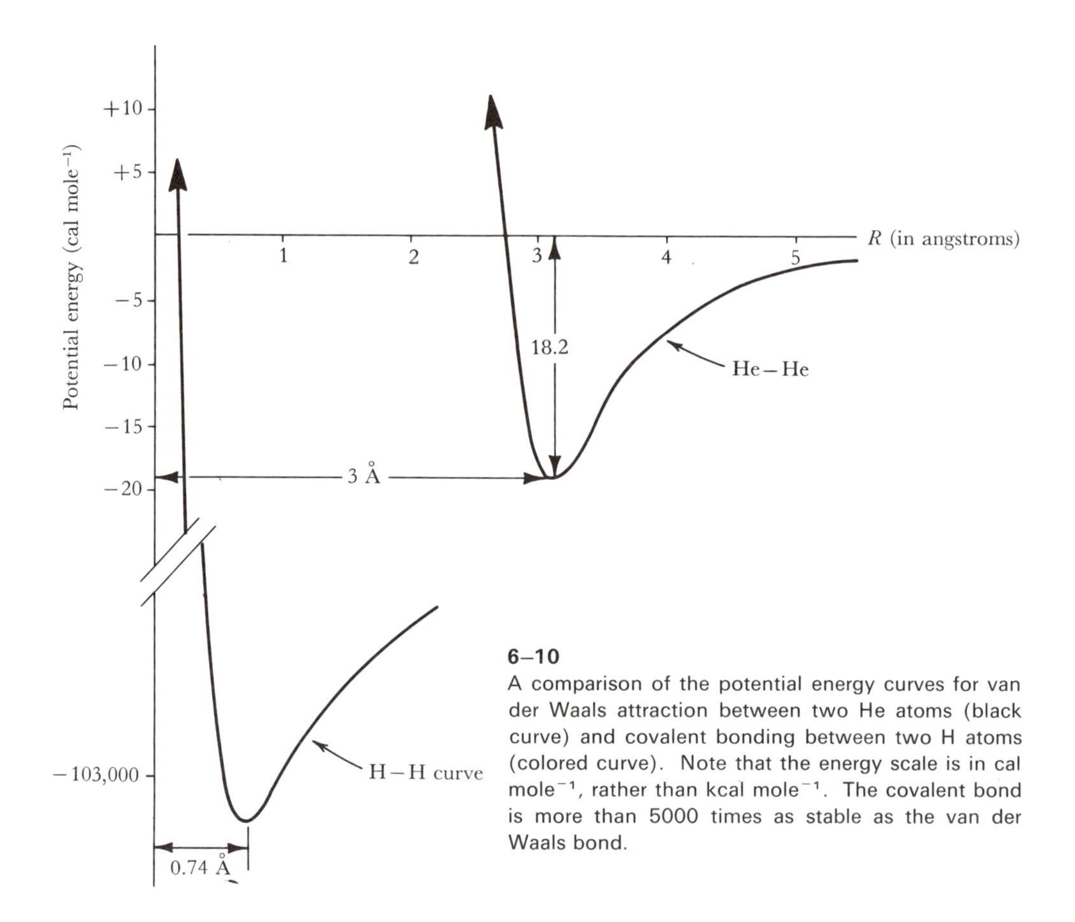

6–10
A comparison of the potential energy curves for van der Waals attraction between two He atoms (black curve) and covalent bonding between two H atoms (colored curve). Note that the energy scale is in cal $mole^{-1}$, rather than kcal $mole^{-1}$. The covalent bond is more than 5000 times as stable as the van der Waals bond.

than the electronic repulsion between the two He atoms. The H–H bond length in the H_2 molecule is 0.74 Å, whereas the equilibrium distance of van der Waals-bonded He atoms is 3 Å. Moreover, a covalent bond is much stronger than a weak van der Waals interaction. Only 18.2 cal $mole^{-1}$ is required to separate helium atoms at their equilibrium distance, but 103,000 cal $mole^{-1}$ is needed to break the covalent bond in H_2.

Molecular solids, in which only van der Waals intermolecular bonding exists, generally melt at low temperatures. This is because relatively little energy of thermal motion is needed to overcome the energy of van der Waals bonding. The liquid and solid phases of helium, which result from weak van der Waals "bonds," exist only at temperatures below 4.6°K. Even at temperatures

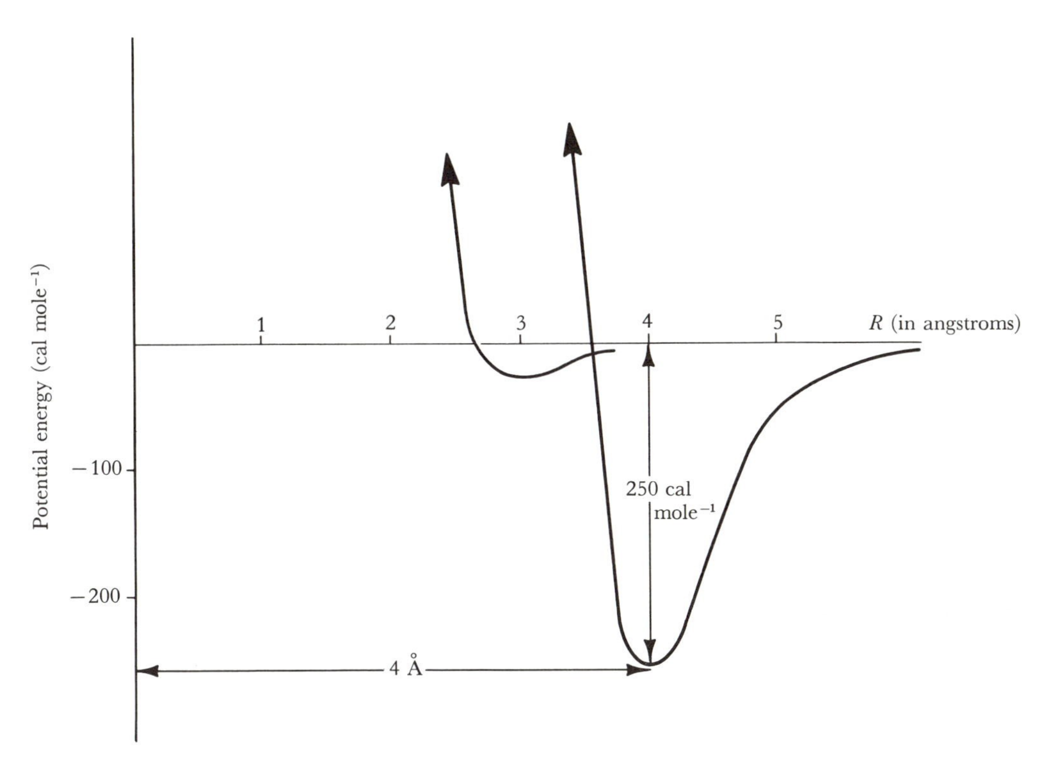

6–11
A comparison of the potential energy curves for van der Waals attraction between two Ar atoms (black curve) and two He atoms (colored curve). The larger Ar atoms are more tightly held, although the bond energy is still one four-hundredth that of a H—H bond.

near absolute zero, solid helium can be produced only at high pressures (29.6 atm at 1.76°K).

Van der Waals bonds in molecular solids and liquids generally are stronger with increasing size of the atoms and molecules involved. For example, as the atomic number of the noble gases increases, the strength of the van der Waals bonding increases also, as shown by the Ar–Ar potential energy curve in Figure 6–11. The attraction between the heavier atoms is stronger, presumably because the outer electrons are held more loosely, and larger instantaneous dipoles and induced dipoles are possible. Because of this stronger van der Waals bonding, solid argon melts at −184°C, or 89°K, which is considerably higher than solid helium.

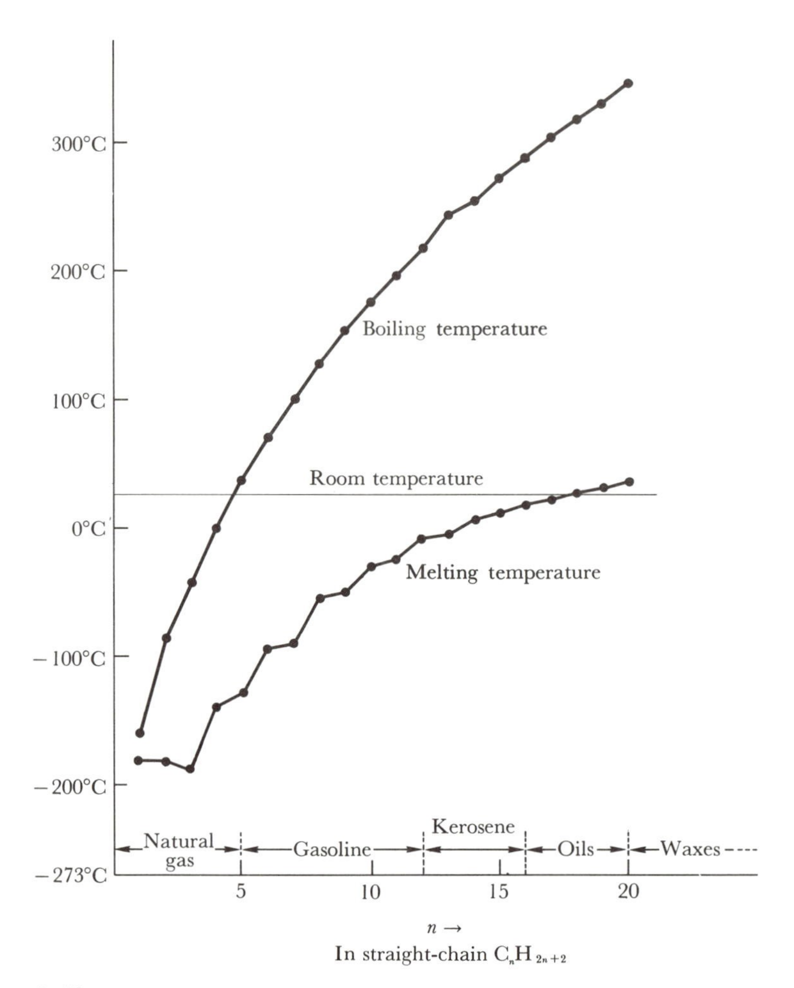

6–12
Melting and boiling temperatures of the straight-chain hydrocarbons as a function of the length of the carbon chain. More energy is required to separate two molecules of eicosane (20 carbons) than ethane (two carbons) because of the more numerous van der Waals interactions between the two larger molecules.

An example of the effect of molecular size on melting and boiling temperatures is provided by a series of straight-chain alkanes, with formulas C_nH_{2n+2}, depicted in Figure 6–12 for $n = 1$ through 20. Part of the increase in melting and boiling temperatures with increasing molecular size and weight arises from the greater energy needed to move a heavy molecule. However, another important factor is the large surface area of a molecule such as eicosane ($C_{20}H_{42}$),

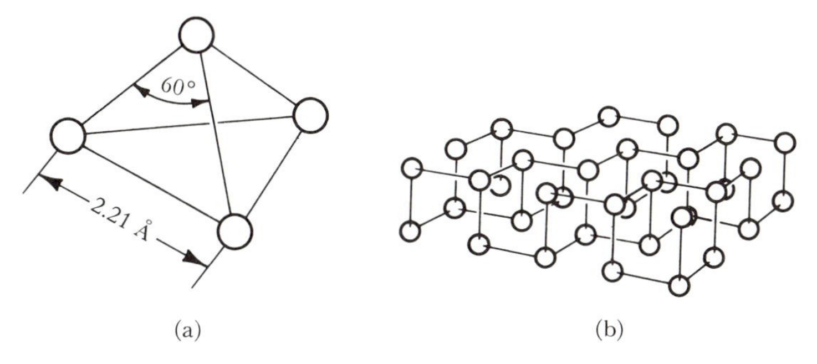

6–13
Structures of solid phosphorus. (a) White phosphorus consists of discrete P_4 molecules. (b) Black phosphorus, a more stable allotrope of the element, has an infinite network structure.

6–14
Structure of solid sulfur. The S_8 ring structure shown is the form sulfur atoms take in the two principal allotropes of crystalline sulfur—rhombic and monoclinic. Rhombohedral sulfur, a third allotrope that is less stable, consists of S_6 rings. A fourth allotrope, amorphous sulfur, contains helical chains of S atoms.

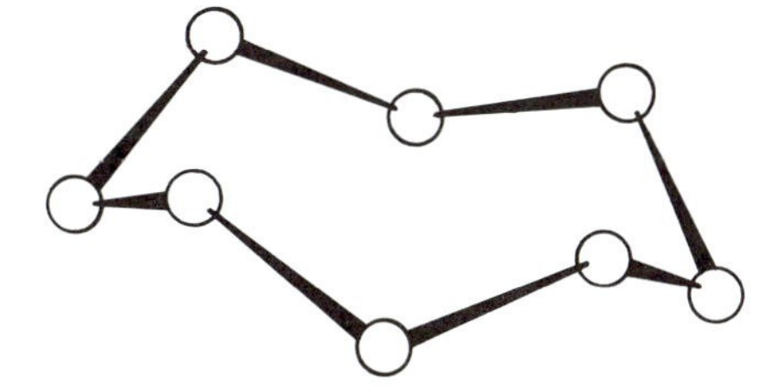

compared with methane, and the greater stability that eicosane therefore can gain from intermolecular van der Waals attractions. The mass effect is similar for both melting and boiling temperatures. However, molecular surface area affects the boiling temperatures more because molecules in the liquid phase are still close enough to exert van der Waals attractions. In fact, without these attractions, which are broken during vaporization, the liquid state could not exist.

P_4 and S_8

Phosphorus exists either as a molecular solid that consists of P_4 molecules or as an infinite network structure (Figure 6–13). The four phosphorus atoms in a P_4 molecule define a regular tetrahedron. Crystalline sulfur consists of discrete S_8 molecules, which have the cyclic structure shown in Figure 6–14. The intermolecular van der Waals bonding is much stronger in solid P_4 than in solid N_2, for example, as indicated by the relatively high melting temperature of solid P_4 (44.1°C), compared to that of N_2 (−210°C). The larger size of the P_4 molecule, and the fact that the outer electrons in P_4 are held much less tightly than in N_2, are the probable explanations of this phenomenon. A similar comparsion can be made between O_2 and S_8. Solid O_2 melts at −219°C, whereas solid S_8 melts at 119°C.

We could ask why solid phosphorus and solid sulfur are not composed of discrete P_2 and S_2 molecules, since these simple diatomic molecules have all their valence orbitals filled, analogous to N_2 and O_2. For the larger atoms it appears that structures in which each atom forms only single bonds with other atoms as in

$$\ddot{\mathrm{P}} \text{ (with three single bonds)} \quad \text{or} \quad -\ddot{\underset{\cdot\cdot}{\mathrm{S}}}-$$

are more stable than structures in which atoms have higher bond orders, hence lower coordination numbers, as in

$$:\mathrm{N}{\equiv}\mathrm{N}: \quad \text{or} \quad :\ddot{\mathrm{O}}{=}\ddot{\mathrm{O}}:$$

At large internuclear separations the strength of π bonding between p orbitals on adjacent atoms is small compared with σ bonding, partly because of a much smaller overlap of the two combining $p(\pi)$ valence orbitals. For example, there is good reason to assume that the σ bonds contribute more to the total bond energies in the P_2 and S_2 molecules than they do in the N_2 and O_2 molecules. Thus one less π bond in S_2 than in P_2 results in only a 11.4% decrease in total bond energy (114 kcal mole^{-1} for P_2 to 101 kcal mole^{-1} for S_2), whereas from N_2 to O_2 a 48% decrease in bond energy (225 kcal mole^{-1} for N_2 to 118 kcal mole^{-1} for O_2) is observed. The relatively weak π bonding in diatomic molecules composed of the larger atoms discourages the formation of structures with higher bond orders. Large atoms of nonmetallic elements tend to form molecular structures that contain only single (σ) bonds.

Polar molecules and hydrogen bonds

Polar molecules are stabilized in a molecular solid by the attractive interaction of oppositely charged ends of the molecules (Figure 6–15). This is called *dipole-dipole interaction*. A particularly important kind of polar interaction is the *hydrogen bond*. This is a bond, which is primarily electrostatic, between a positively charged hydrogen atom and a small, electronegative atom, usually N, O, or F. For example, glycine molecules are held in a sheet structure by van der Waals forces and hydrogen bonds (Figure 6–16). Ice provides another example of the importance of hydrogen bonding in building intermolecular structures. As shown in Figure 6–17, each oxygen atom of a polar H_2O molecule is tetrahedrally coordinated to four other oxygen atoms in a structure that somewhat resembles the diamond structure. Each oxygen atom is bound to its four neighbor oxygen atoms by hydrogen bonds. In two of these hydrogen bonds the central H_2O molecule supplies the hydrogen atoms; in the other two bonds the hydrogen atoms come from neighboring water molecules. Such bonds are weak compared with covalent bonds. A typical covalent bond energy is about 100 kcal mole^{-1}, whereas a hydrogen bond between H and O is approximately 5 kcal mole^{-1}. But hydrogen bonds are important for the

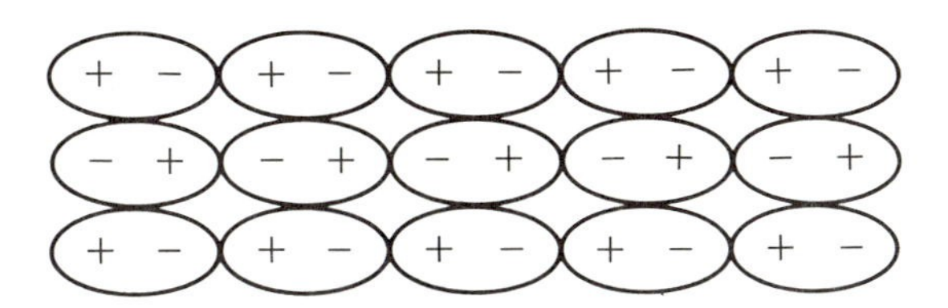

6–15
Diagrammatic representation of the packing of polar molecules into a crystalline solid. Packing occurs so partial charges of opposite sign are in close proximity.

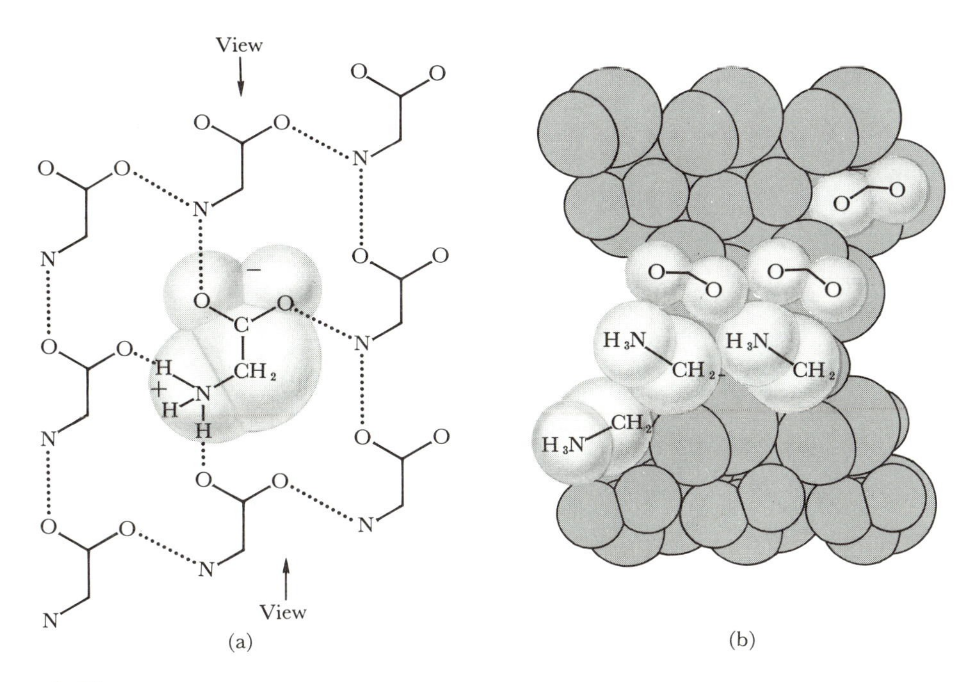

6–16
The bonding in solid glycine, $^{+}H_3N—CH_2—COO^{-}$. (a) Molecules in a layer are packed tightly and are held together by van der Waals attraction and by hydrogen bonds (dotted). (b) The layers are stacked on top of one another and held together by van der Waals attractions. With this perspective the layers are on edge, and in a horizontal position. The view of the layers in (b) is marked by arrows in (a).

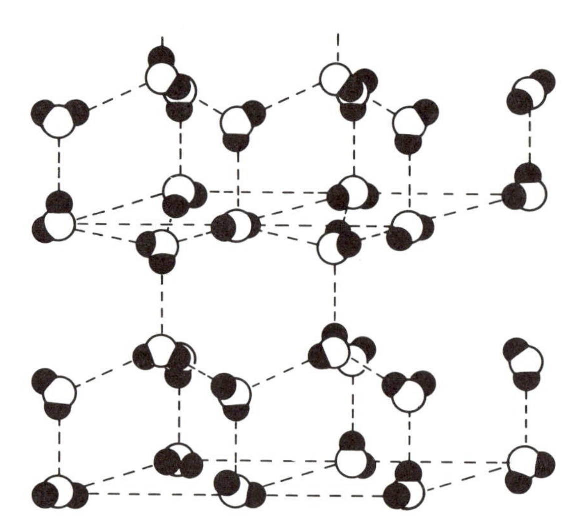

6–17
In crystalline ice each oxygen atom is hydrogen bonded to two others by means of its own hydrogen atoms, and bonded to two more oxygen atoms by means of their hydrogen atoms. The coordination is tetrahedral, and the structure is similar to that of diamond.

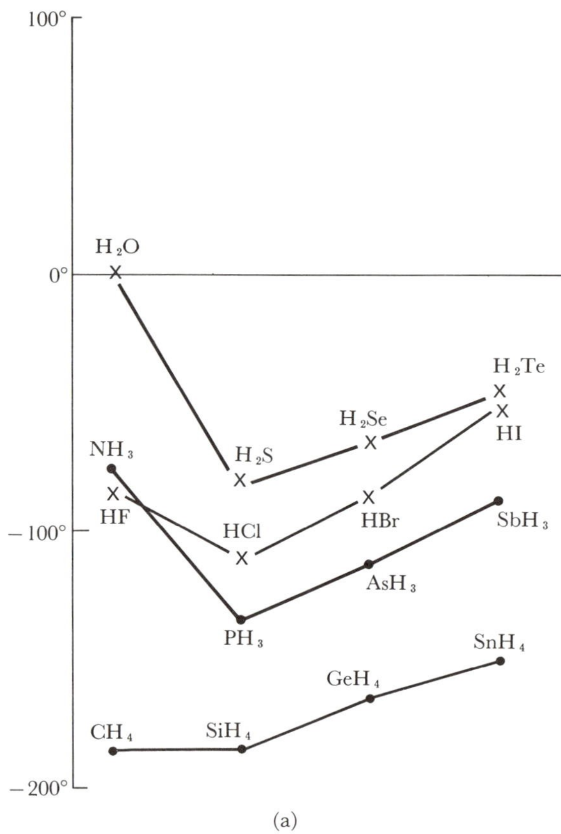

(a)

same reason that van der Waals bonds are: They may be weak but there are many of them.

Hydrogen bonding in water is responsible for many of its most important properties. Because of hydrogen bonds in both the solid and liquid phases the melting and boiling temperatures of water are unexpectedly high when compared with those of H_2S, H_2Se, and H_2Te, which are hydrogen compounds of elements also in Group VIA of the periodic table. Solid and liquid ammonia and hydrogen fluoride show anomalous behavior similar to water and for the same reason (Figure 6–18). However, hydrogen bonding in ammonia is less pronounced than in water for two reasons: N is less electronegative than O, and NH_3 has only one lone pair of electrons to attract the H from a neighboring molecule. In contrast, HF is less well hydrogen-bonded than is H_2O in spite of the greater electronegativity of F and the presence of three lone pairs. This is because HF has only one hydrogen atom to use in making such bonds.

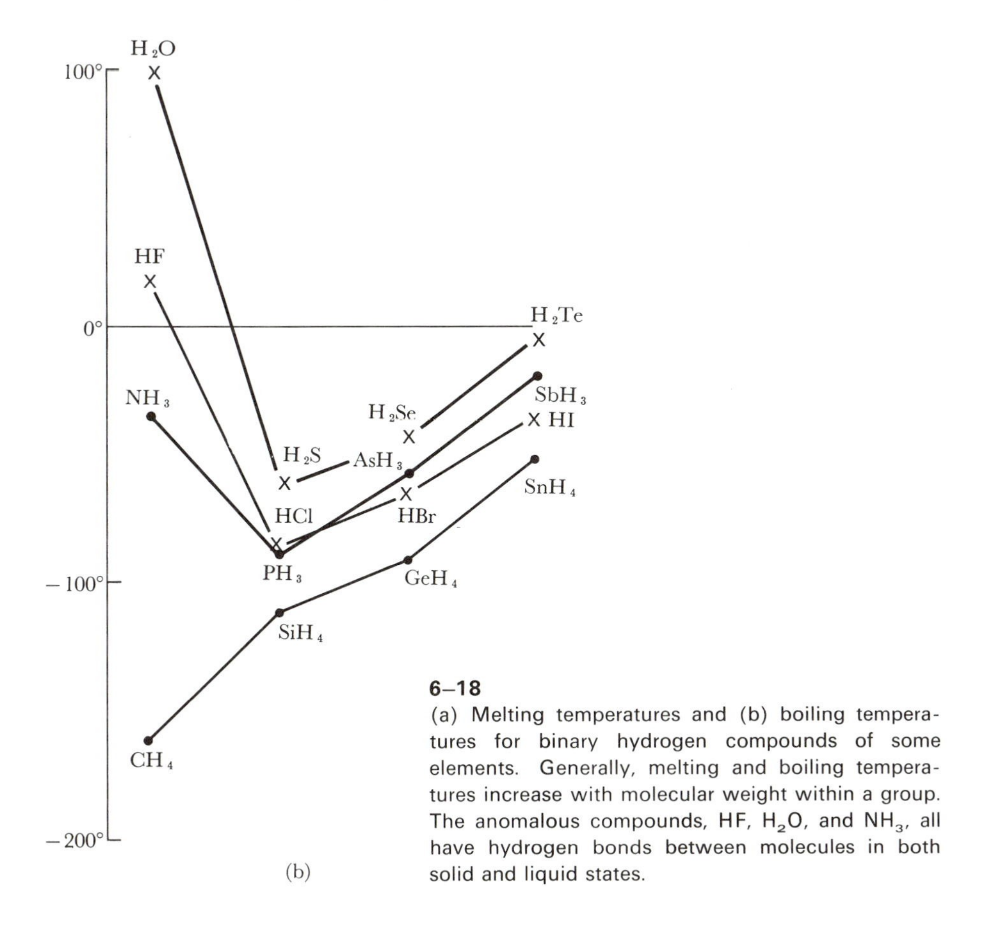

6–18
(a) Melting temperatures and (b) boiling temperatures for binary hydrogen compounds of some elements. Generally, melting and boiling temperatures increase with molecular weight within a group. The anomalous compounds, HF, H_2O, and NH_3, all have hydrogen bonds between molecules in both solid and liquid states.

Since hydrogen bonding causes an open network structure in ice (Figure 6–17), ice is less dense than water at the melting temperature. Upon melting, part of this open-cage structure collapses, and the liquid is more compact than the solid. The measured heat of fusion of ice is only 1.4 kcal mole^{-1}, whereas the energy of its hydrogen bonds is 5 kcal mole^{-1}. This indicates that only about 28% of the hydrogen bonds of ice are broken when it melts. Water is not composed of isolated, unbonded molecules of H_2O; rather, it has regions or clusters of hydrogen-bonded molecules. That is, part of the hydrogen-bonded structure of the solid persists in the liquid. As the temperature is raised these clusters break up, and the volume continues to shrink. If the temperature is raised still higher, the expected thermal expansion dominates over the shrinkage caused by the collapse of the cage structures. Consequently liquid water has a minimum molar volume, or a maximum density, at 4°C.

Because the hydrogen-bonded H_2O clusters are broken slowly as heat is added, water has a higher specific heat than any other common liquid except ammonia. Water also has an unusually high heat of fusion and heat of vaporization. All three of these properties give water the capability to act as a large thermostat, which confines the temperature on the earth within moderate limits. Ice absorbs a large amount of heat when it melts, and water absorbs more heat per unit of temperature rise than almost any other substance. Correspondingly, as water cools, it gives off more heat to its surroundings than other substances.

Polar molecules as solvents

The polar nature of liquid water makes it an excellent solvent for ionic solids such as NaCl. Water can dissolve NaCl and separate the oppositely charged Na^+ and Cl^- ions because the energy required to separate the ions is provided by the formation of hydrated ions (Figure 6–19). Each Na^+ ion in solution still has an octahedron of negative charges around it, but instead of being Cl^- ions they are the negative poles of the oxygen atoms of the water molecules. The Cl^- ions also are hydrated, but it is the positive end (H) of the water molecules that approach the Cl^- ions. A nonpolar solvent such as gasoline, a liquid composed of hydrocarbon molecules, cannot form such *ion-dipole bonds* with Na^+ and Cl^-. Consequently NaCl and other salts are insoluble in gasoline.

Polar solvents dissolve polar molecular solids because of dipole-dipole interactions. The energy released by the formation of dipole-dipole bonds between a polar solvent and solute molecules is sufficient to break the intermolecular forces in the molecular solids (Figure 6–20). For example, ice is soluble in liquid ammonia but not in benzene because NH_3 is a polar molecule, whereas C_6H_6 is nonpolar.

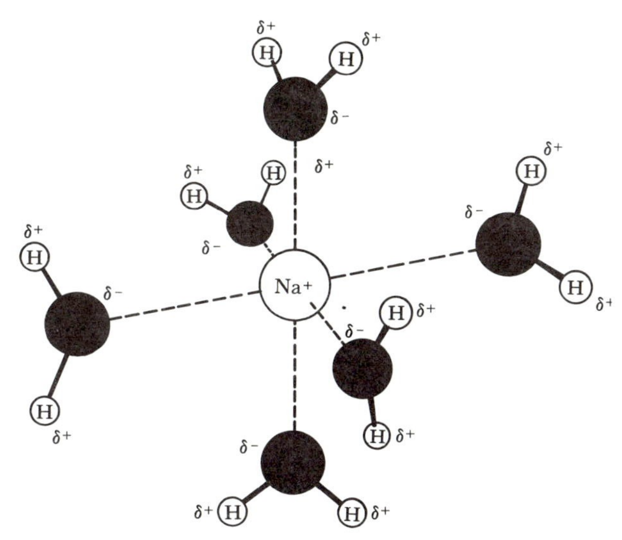

6–19
In solution, the hydrated Na^+ ion is surrounded by an octahedron of negative charges, but these negative charges are from the dipolar solvent molecule, H_2O, instead of the Cl^-.

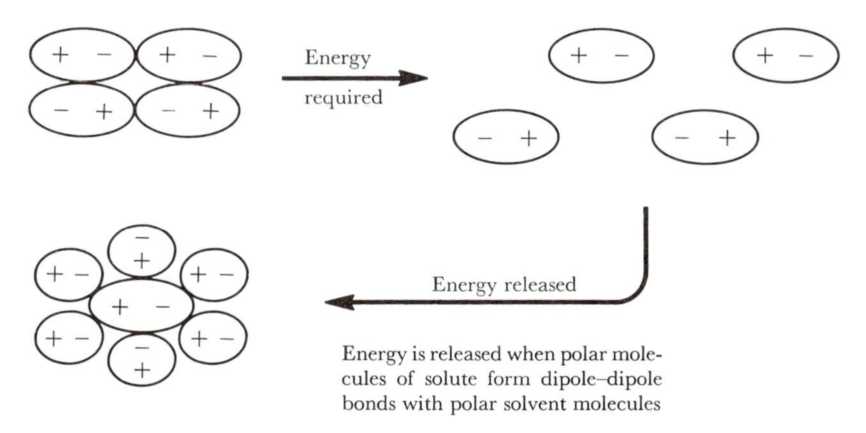

6–20
When a crystalline solid composed of polar molecules dissolves, stability is lost when oppositely charged ends of neighboring molecules are removed. This loss is compensated by the stability produced by solvating the polar molecules in solution. A solvent that cannot provide such stabilization cannot dissolve the solid.

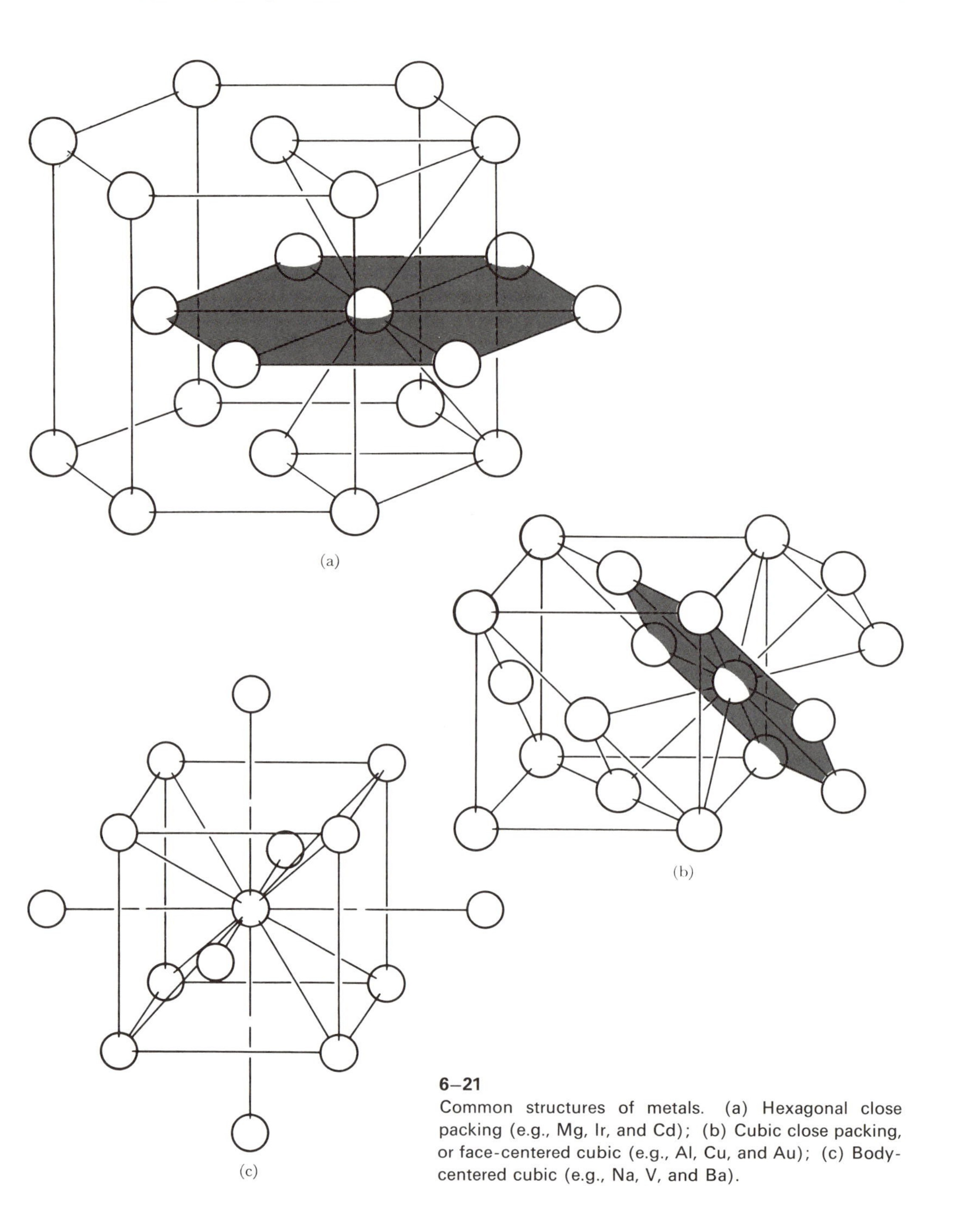

6–21
Common structures of metals. (a) Hexagonal close packing (e.g., Mg, Ir, and Cd); (b) Cubic close packing, or face-centered cubic (e.g., Al, Cu, and Au); (c) Body-centered cubic (e.g., Na, V, and Ba).

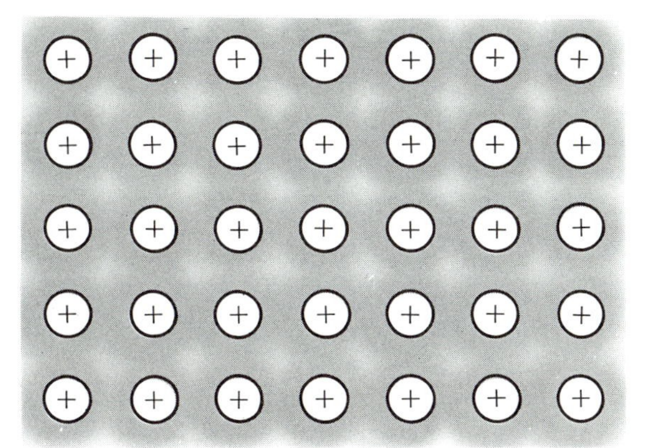

6–22
Cross section of a crystal structure of a metal with the sea of electrons. Each circled positive charge represents the nucleus and filled, nonvalence electron shells of a metal atom. The shaded area surrounding the positive metal ions indicates schematically the mobile sea of electrons.

6–3 METALS

In metals at least eight "nearest-neighbor" atoms surround a particular atom in one of the three common structures shown in Figure 6–21. In both hexagonal and cubic close packing, each sphere touches 12 other spheres, six in a plane, three above, and three below. X-ray analysis reveals that two thirds of all metals crystallize in one of these two structures. A majority of the other one third crystallize as body-centered cubes, in which each atom has only eight nearest neighbors.

The properties of metals suggest that the valence electrons are relatively free to move through the crystal structure. Figure 6–22 illustrates one model in which the electrons form a sea of negative charges that holds the atoms tightly together. The circled positive charges represent the positively charged ions remaining when valence electrons are stripped away, leaving the nuclei and the filled electron shells. Since metals generally have high melting temperatures and high densities, especially in comparison with molecular solids, the "electron sea" must strongly bind the positive ions in the crystal.

The simple electron-sea model for metallic bonding also is consistent with two other commonly observed properties of metals: malleability and ductility. A malleable material can be hammered easily into sheets; a ductile material can be drawn into thin wires. For metals to be shaped and drawn without fracturing, the atoms in the planes of the crystal structure must be displaced easily with respect to each other. This displacement does not result in the

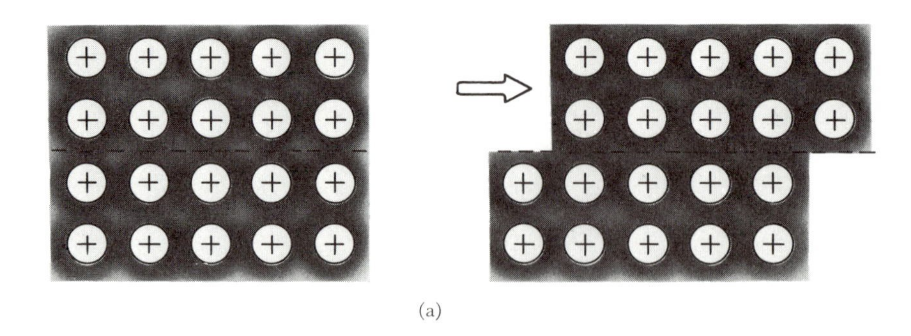

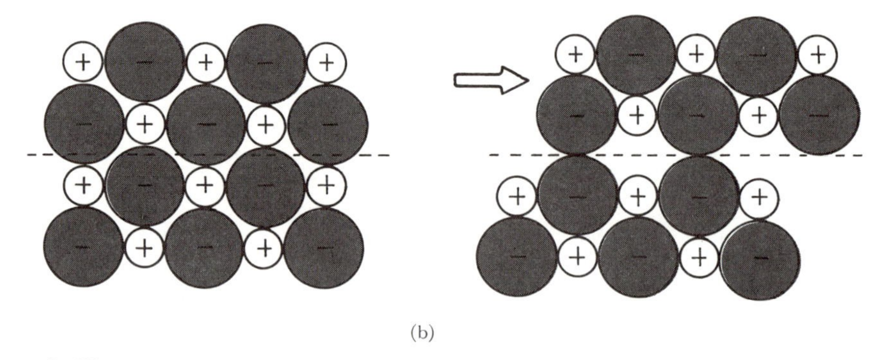

6–23
(a) Shift of metallic crystal along a plane results in no strong repulsive forces. (b) Shift of ionic crystal along a plane results in strong repulsive forces and crystal distortion.

development of strong repulsive forces in metals because the mobile sea of electrons provides a constant buffer, or shield, between the positive ions. This situation is in direct contrast to ionic crystals, in which the binding forces are due almost entirely to electrostatic attractions between oppositely charged ions. In an ionic crystal valence electrons are bound firmly to the atomic nuclei. Displacement of layers of ions in such a crystal brings ions of like charge together and causes strong repulsions that can lead to crystal fracture (Figure 6–23).

Electronic bands in metals

The delocalized molecular orbital theory provides a more detailed (and more informative) model for metallic bonding. In this model the entire block of

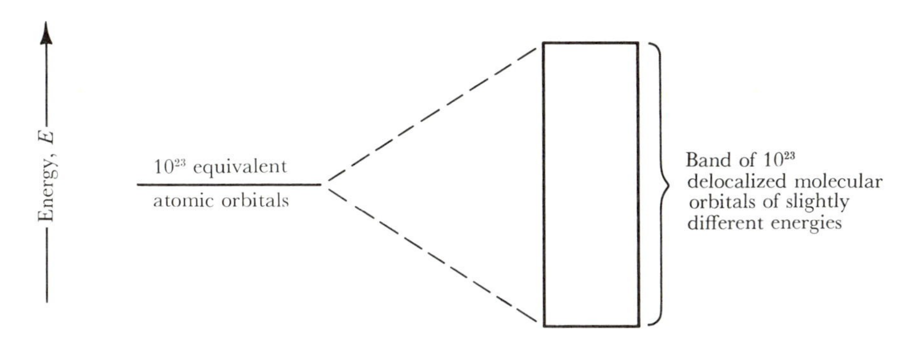

6–24
Just as six atomic p orbitals can combine to produce six delocalized molecular orbitals in benzene, 10^{23} atomic orbitals of a metal can combine to produce 10^{23} metallic orbitals. The metallic orbitals are so closely spaced that they can be treated as a continuous band of energy. The simple band theory of metals can explain many of their properties.

metal is considered as a giant molecule. All the atomic orbitals of a particular type in the crystal interact to form a set of delocalized orbitals that extend throughout the entire block. For a particular crystal, assume that the number of valence orbitals is of the order of 10^{23}. Figure 6–24 depicts the combination of approximately 10^{23} equivalent atomic orbitals in a crystal to form 10^{23} delocalized orbitals. All of these orbitals cannot have the same energy if they are delocalized. However, instead of producing an antibonding orbital and a bonding molecular orbital, as a diatomic molecule does, the combination of atomic orbitals produces a band of closely spaced energy levels.

Figure 6–25 illustrates the three bands of energy levels formed by the 1*s*, 2*s*, and 2*p* orbitals of the simplest metal, lithium. The 1*s* molecular orbitals are filled completely because the 1*s* atomic orbitals in isolated lithium atoms are filled. Thus the 1*s* electrons make no contribution to bonding. They are part of the positive ion cores and can be eliminated from the discussion. Atomic lithium has one valence electron in a 2*s* orbital. If there are 10^{23} atoms in a lithium crystal, the 10^{23} 2*s* orbitals interact to form a band of 10^{23} delocalized orbitals. As usual, each of these orbitals can accommodate two electrons, so the capacity of the band is 2×10^{23} electrons. Lithium metal has enough electrons to fill only the lower half of the 2*s* band, as illustrated in Figure 6–25.

The presence of a partially filled band of delocalized orbitals accounts for bonding and electrical conduction in metals. Electrons in the lower filled orbital band move throughout the crystal in a random fashion such that their motion results in no *net* separation of electrons and positive ions in the metal. For a metal to conduct an electric current, electrons must be excited to unfilled delocalized orbitals in such a way that their movement in one direction is not

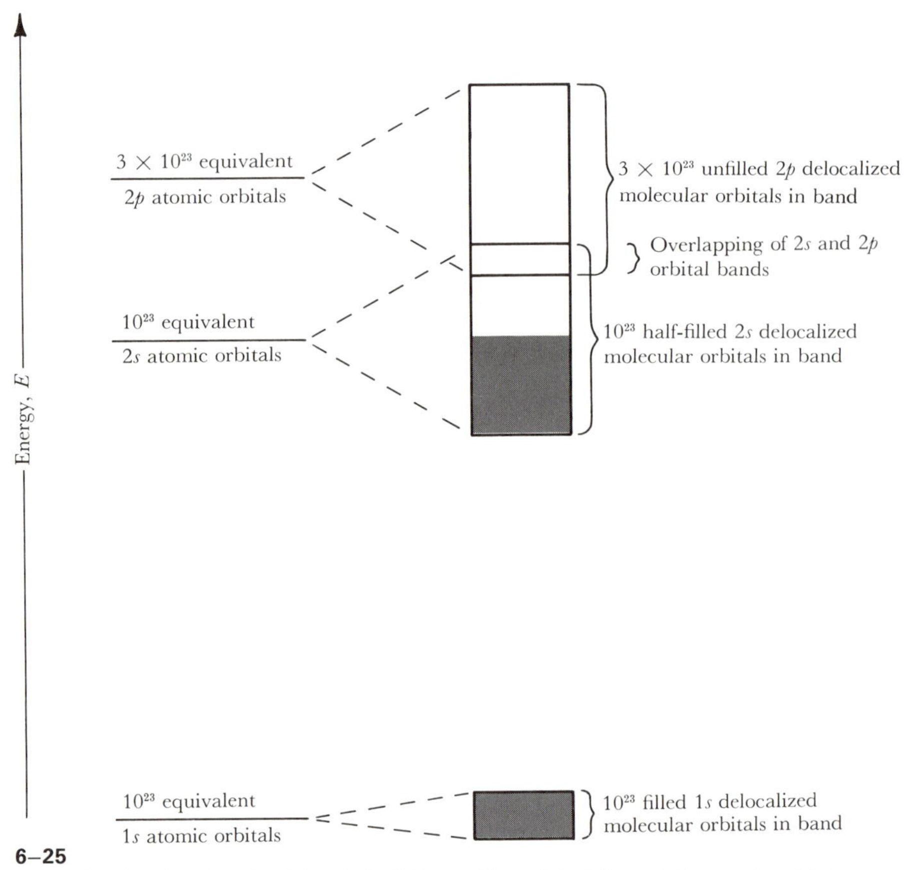

6–25
Delocalized molecular orbital bands in lithium. The original $2s$ and $2p$ atomic orbitals are so close in energy that the molecular orbital bands overlap. Lithium has one electron in every $2s$ atomic orbital, hence only half as many electrons as can be accommodated in the $2s$ atomic orbitals or in the delocalized molecular orbital band. There are unfilled energy states an infinitesimal distance above the highest-energy filled state, so an infinitesimal energy is required to excite an electron and send it moving through the metal. Thus lithium is a conductor.

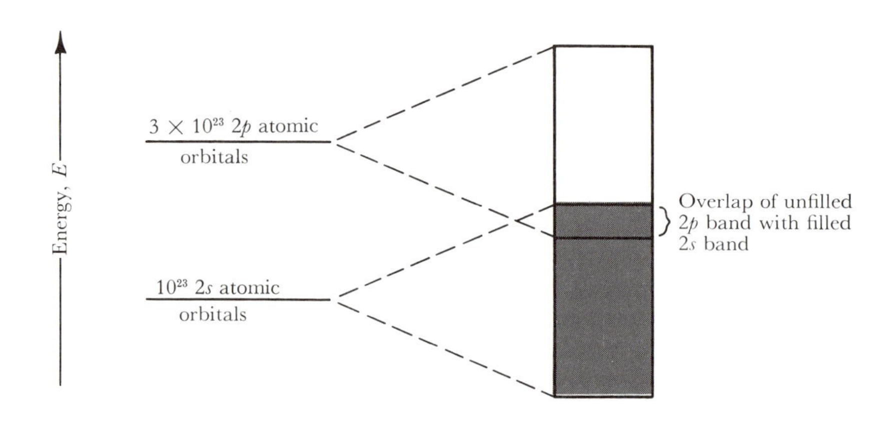

exactly canceled by electrons moving in the opposite direction. Such concerted electron movement occurs only when an electric potential difference is applied between two regions of a metal. Then electrons are excited to the unfilled delocalized molecular orbitals that are part of the same band (the 2*s* band for lithium) and just slightly higher in energy. Therefore we can expect a metal to conduct electricity. Conduction is restricted by the frequent collisions of electrons with the positive ions, which have kinetic energy and thus vibrate randomly within their crystal sites. As the temperature increases, vibration of the positive ions increases, and collisions with the conduction electrons are more frequent. Therefore electrical conductivity in metals decreases as the temperature increases.

Beryllium is a more complicated example than lithium. An isolated beryllium atom has exactly enough electrons to fill its 2*s* orbital. Accordingly, beryllium metal has enough electrons to fill its 2*s* delocalized band. If the 2*p* band did not overlap the 2*s* (Figure 6–26), beryllium would not conduct well because an energy equal to the gap between bands would be required before electrons could move through the solid. However, the two bands do overlap and beryllium has unoccupied delocalized orbitals that are an infinitesimal distance above the most energetic filled orbitals. Consequently beryllium is a metallic conductor.

6–4 NONMETALLIC NETWORK SOLIDS

Nonmetallic network materials such as boron or carbon are *insulators*; that is, they do not conduct electrical current. One way to visualize the difference between nonmetallic insulators and metallic conductors is to use approximate localized orbitals to describe the structure of insulators. We can use localized bonds to describe insulators quite accurately because their atomic coordination numbers are relatively low. Because of the low coordination numbers there usually are enough electrons in the valence orbitals to form three or four simple covalent bonds between each atom and its nearest neighbors. The construction of these bonds resembles the formation of localized bonds in a polyatomic molecule.

◀ **6–26**
Band-filling diagram for beryllium. A Be atom has enough electrons (two) to fill its 2*s* orbital, so Be metal has enough electrons to fill its 2*s* delocalized molecular-orbital band. If the 2*s* and 2*p* bands did not overlap, Be would be an insulator because an appreciable amount of energy would be required to make electrons flow in the solid. But with the band overlap shown here, an infinitesimal amount of energy excites electrons to the 2*p* band orbitals and electrons flow.

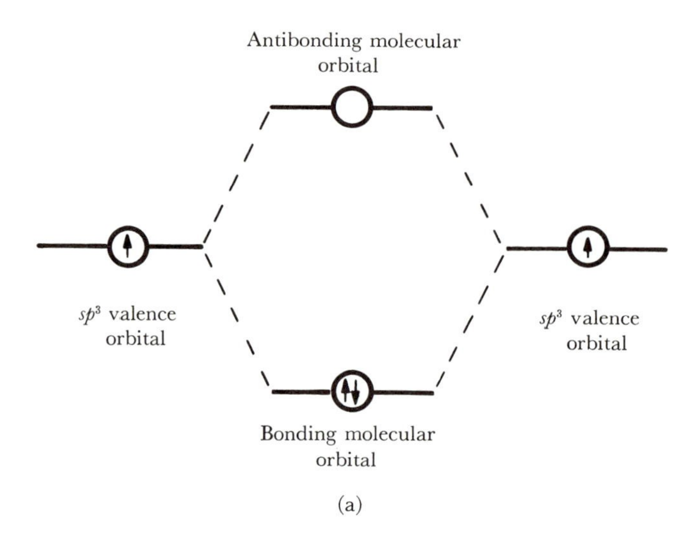

(a)

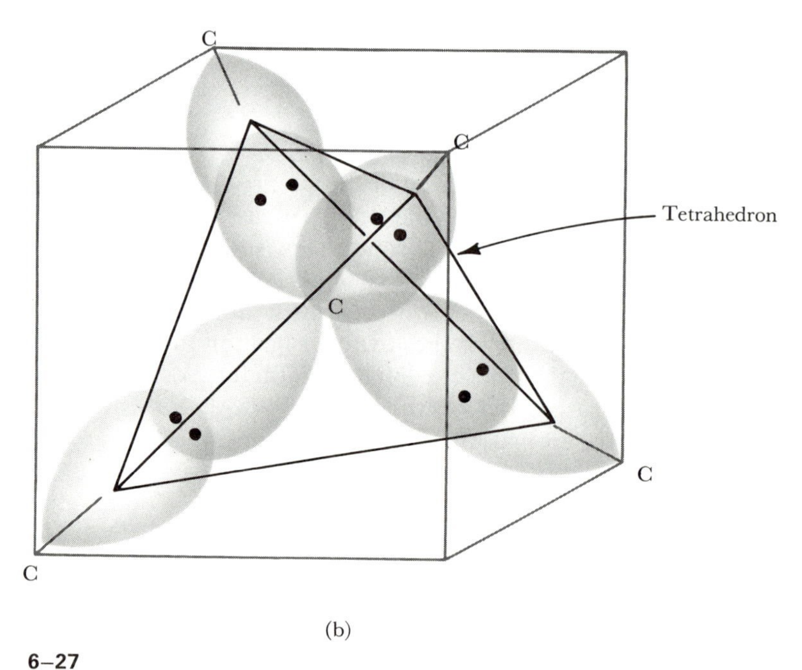

(b)

6–27
Bonding in diamond crystals. (a) Localized-orbital energy levels in diamond crystals. Each pair of neighboring localized atomic sp^3 orbitals produces a bonding orbital and an antibonding orbital. (b) Schematic representation of the overlap of the four sp^3 hybrid orbitals of a C atom with similar orbitals from four other carbon atoms.

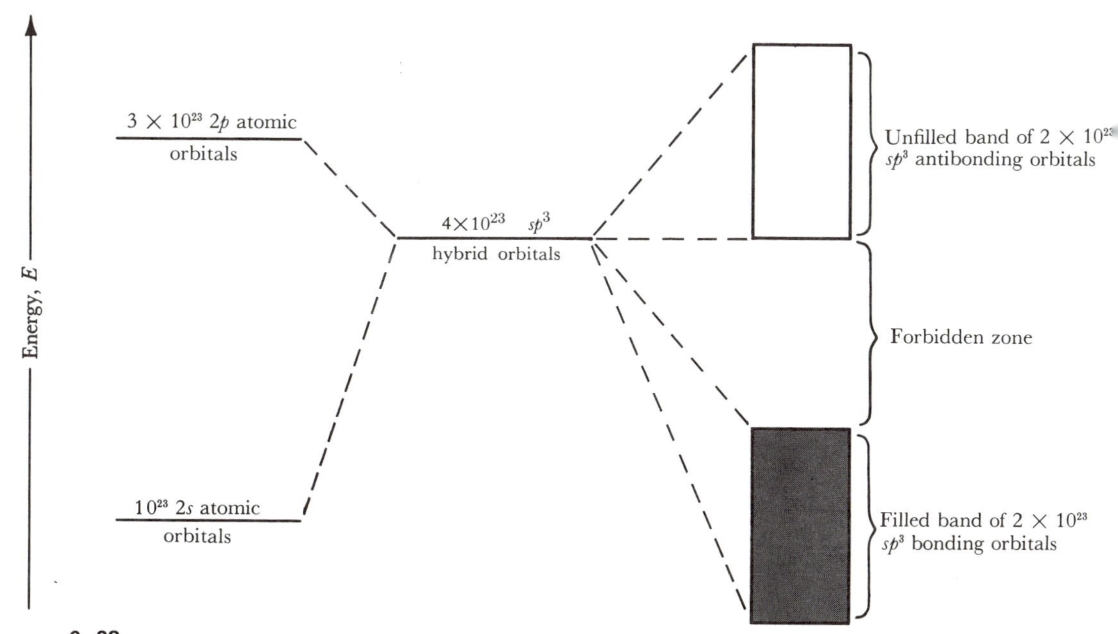

6–28
Delocalized molecular orbital bands in an insulator, formed from equivalent sp^3 localized hybrid orbitals. Note the relatively large zone between the filled band of sp^3 bonding orbitals and the unfilled band of sp^3 antibonding orbitals.

For diamond we begin to construct the bonding model by assigning each carbon atom four localized tetrahedral sp^3 hybrid orbitals. One such orbital from each of two neighboring carbon atoms combine to make one bonding and one antibonding molecular orbital (Figure 6–27). The four valence electrons in each carbon atom are sufficient to fill these bonding orbitals. Thus all electrons in diamond are used for bonding, thereby leaving none to move freely to conduct electricity.

To construct the band model of delocalized orbitals for an insulating network solid such as diamond, we will proceed as follows. Assume 10^{23} carbon atoms. When the 4×10^{23} localized sp^3 orbitals interact with each other, two bands of delocalized orbitals are formed, one from the 2×10^{23} bonding orbitals of Figure 6–27 and one from the 2×10^{23} antibonding orbitals. These are depicted in Figure 6–28, in which the atomic orbitals are drawn at the left to remind you that these orbitals originally came from the 2*s* and 2*p* atomic orbitals. The important fact in this diagram is that the band filled with electrons does not overlap with the next higher energy band, which has completely unfilled orbitals. There is a forbidden energy zone or gap between what is called the valence band below and the conduction band above. There are 4×10^{23} valence electrons per 10^{23} carbon atoms, enough to fill completely the 2×10^{23} orbitals in the valence band.

For an insulator to conduct, energy is required that is sufficient to excite electrons in the filled band across this forbidden energy zone into the unfilled molecular orbitals. This energy is the activation energy of the conduction process. Only high temperatures or extremely strong electrical fields will provide enough energy to an appreciable number of electrons for conduction to occur. In diamond the gap between the top of the valence band and the bottom of the conduction band is 5.2 eV, or 120 kcal $mole^{-1}$.

Semiconductors

The border line between metallic and nonmetallic network structures of elements in the periodic table is not sharp (Figure 6–6). This is shown by the fact that several elemental solids have properties that are intermediate between conductors and insulators. Silicon, germanium, and α-gray tin all have the diamond structure. However, the forbidden energy gap between filled and empty bands for these solids is much smaller than for carbon. Rather than 120 kcal $mole^{-1}$ for carbon, the gap for silicon is only 25 kcal $mole^{-1}$. For germanium it is 14 kcal $mole^{-1}$, and for α-gray tin it is 1.8 kcal $mole^{-1}$. The metalloids silicon and germanium are called *semiconductors*. Figure 6–29 shows the band diagram for a semiconductor, with a small forbidden energy zone.

A semiconductor can carry a current if the relatively small energy required to excite electrons from the lower filled valence band to the upper empty conduction band is provided. Since the number of excited electrons increases as the temperature increases, the conductance of the semiconductor increases with temperature. This behavior is exactly the opposite of that of metals.

Conduction in materials such as silicon and germanium can be enhanced by adding small amounts of certain impurities. Although there is a forbidden energy gap in silicon, it can be narrowed effectively if impurities such as boron or phosphorus are added to silicon crystals. Small amounts of boron or phosphorus (a few parts per million) can be incorporated into the silicon structure when the crystal is grown. Phosphorus has five valence electrons and thus has an extra free electron even after four electrons have been used in the covalent bonds of the silicon structure. This fifth electron can be moved away from a phosphorus atom by an electric field; hence we say phosphorus is an electron donor. Only 0.25 kcal $mole^{-1}$ is required to free the donated electrons, thereby making a conductor out of silicon to which a small amount of phosphorus has been added. The opposite effect occurs if boron instead of phosphorus is added to silicon. Atomic boron has one too few electrons for complete covalent bonding. Thus for each boron atom in the silicon crystal there is a single vacancy in a bonding orbital. It is possible to excite the valence

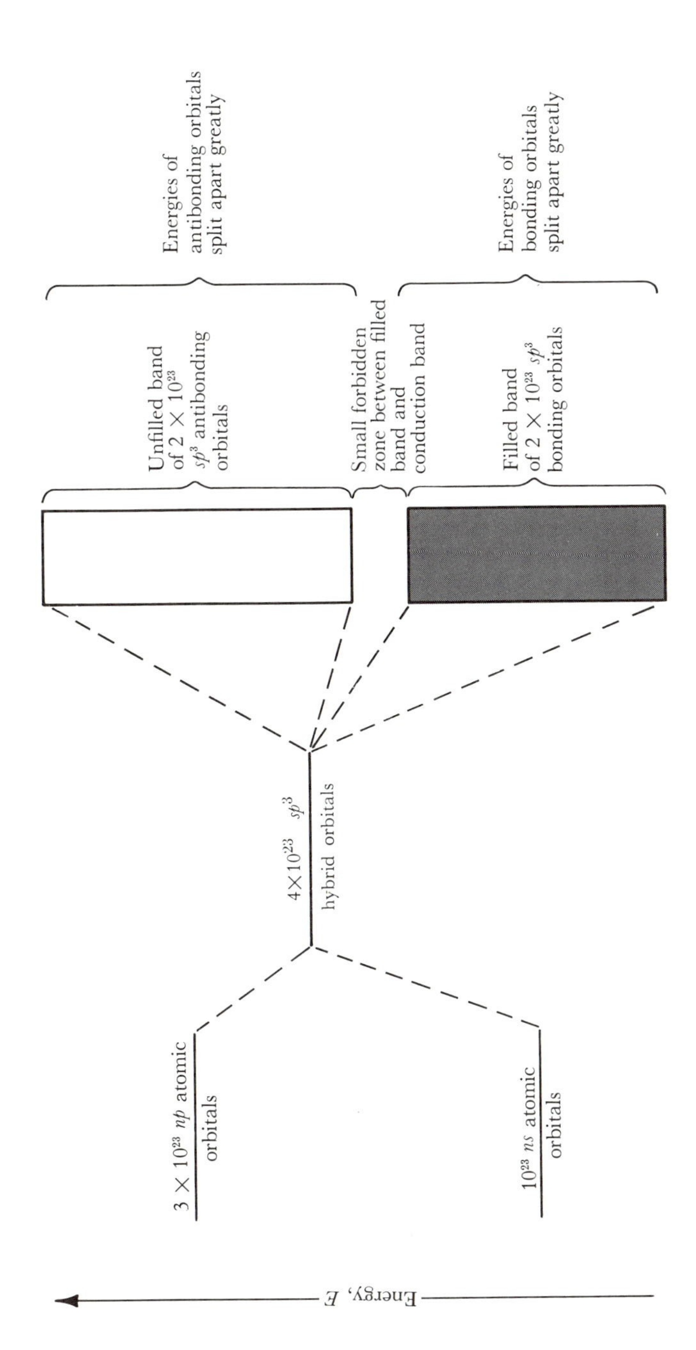

6–29
Bands of delocalized orbitals in semiconductors formed from equivalent sp^3 localized hybrid orbitals. The forbidden zone between filled and empty bands is smaller than in an insulator.

electrons of silicon into these vacant orbitals in the boron atoms, thereby causing the electrons to move through the crystal. To accomplish this conduction an electron from a silicon neighbor drops into the empty boron orbital. Then an electron that is two atoms away can fill the silicon atom's newly created vacancy. The result is a cascade effect, whereby an electron from each of a row of atoms moves one place to the neighboring atom. Physicists prefer to describe this phenomenon as a hole moving in the opposite direction. No matter which description is used, it is a fact that less energy is required to make a material such as silicon conduct if the crystal contains small amounts of either an electron donor such as phosphorus or an electron acceptor such as boron.

Silicates

The earth's crust—the upper 20 miles under the continents and as little as three miles under the ocean beds—consists mainly of silicate minerals. The mantle, a layer about 1800 miles thick beneath the crust, probably is composed of dense silicates. The crust is about 48 % oxygen by weight, 26 % silicon, 8 % aluminum, 5 % iron, and 11 % calcium, sodium, potassium, and magnesium combined.

The basic building block in silicates is the orthosilicate ion, SiO_4^{4-}, shown in Figure 6–30. Each silicon atom is covalently bonded to four oxygen atoms at the corners of a tetrahedron. The SiO_4^{4-} anion occurs in simple minerals such as zircon ($ZrSiO_4$), garnet, and topaz. Two tetrahedra can share a corner oxygen atom to form a discrete $Si_2O_7^{6-}$ anion, or three tetrahedra can form a ring, shown in Figure 6–31. Benitoite, $BaTiSi_3O_9$, is the best known example of this uncommon kind of silicate. Beryl, $Be_3Al_2Si_6O_{18}$, a common source of beryllium, has anions composed of rings of six tetrahedra with six shared oxygen atoms.

Chain structures. All of the silicates mentioned so far are made from discrete anions. A second class is made of endless strands or chains of linked tetrahedra. Some minerals have single silicate strands with the formula $(SiO_3)_n^{2n-}$. A form of asbestos has the double-stranded structure shown in Figure 6–32. The double-stranded chains are held together by electrostatic forces between themselves and the Na^+, Fe^{2+}, and Fe^{3+} cations packed around them. The chains can be pulled apart with much less effort than is required to snap the covalent bonds within a chain. Therefore asbestos has a stringy, fibrous texture. Aluminum ions can replace as many as one quarter of the silicon ions in the tetrahedra. However, each replacement requires one more positive charge from another cation (such as K^+) to balance the charge on the silicate oxygen atoms. The physical properties of the silicate minerals are influenced strongly by how many Al^{3+} ions replace Si^{4+} ions, and by how many extra cations therefore are needed to balance the charge.

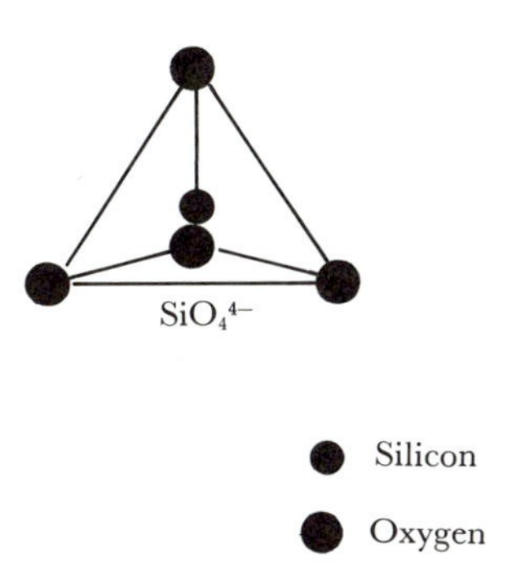

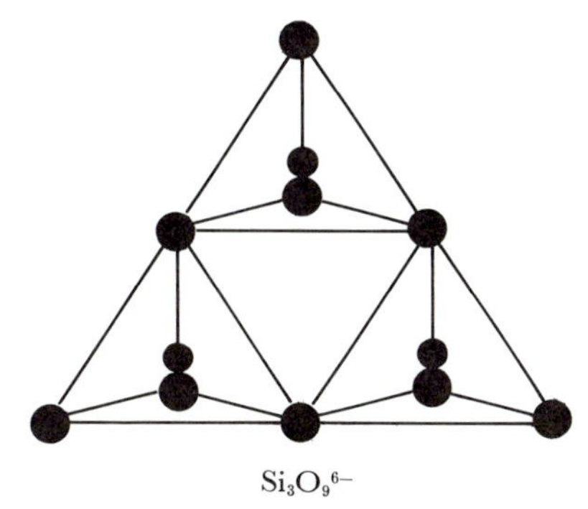

6–30
The SiO_4^{4-} tetrahedron, which is the building block of most silicate minerals. The Si atom (black) is covalently bonded to four oxygen atoms (color) at the corners of a tetrahedron. The covalent bonds are not shown. The black lines between oxygen atoms are included only to give form to the tetrahedron.

6–31
A ring of three tetrahedra, with three oxygen atoms shared between pairs of tetrahedra, has the formula $Si_3O_9^{6-}$. This structure occurs as the anion in benitoite, $BaTiSi_3O_9$.

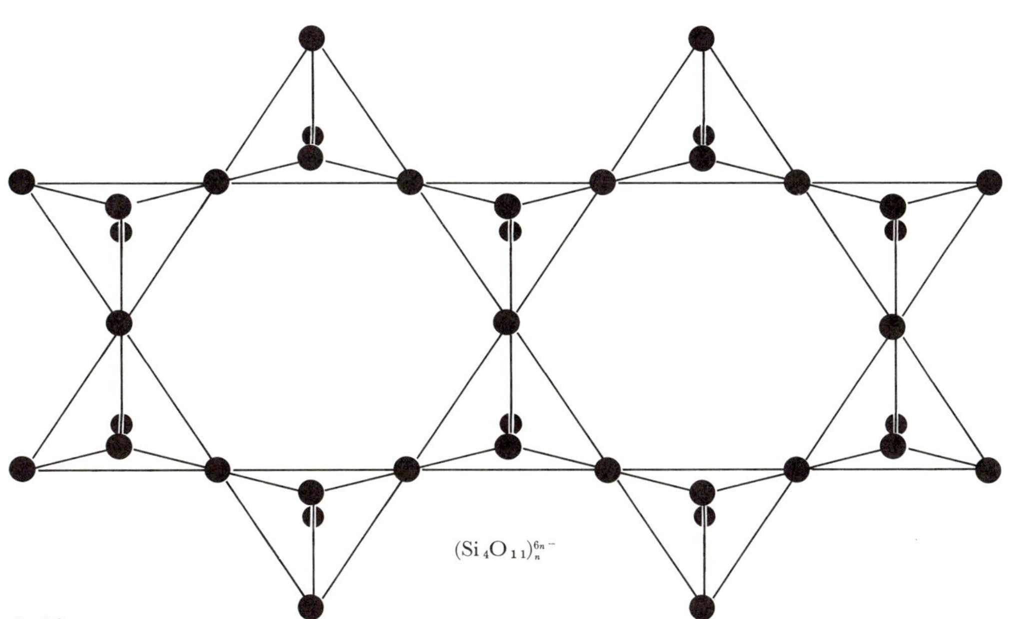

6–32
Long double-stranded chains of silicate tetrahedra are in fibrous minerals such as asbestos. A common natural form of asbestos has the empirical formula $Na_2Fe_3^{2+}Fe_2^{3+}(Si_8O_{22})(OH)_2$.

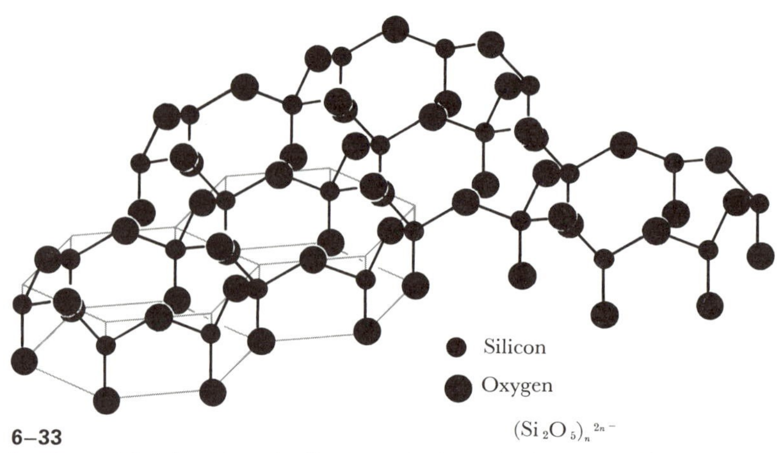

6–33
In mica and the clay minerals silicate tetrahedra each share three of their corner oxygen atoms to make endless sheets. All of the unshared oxygen atoms point down in this drawing on the same side of the sheet.

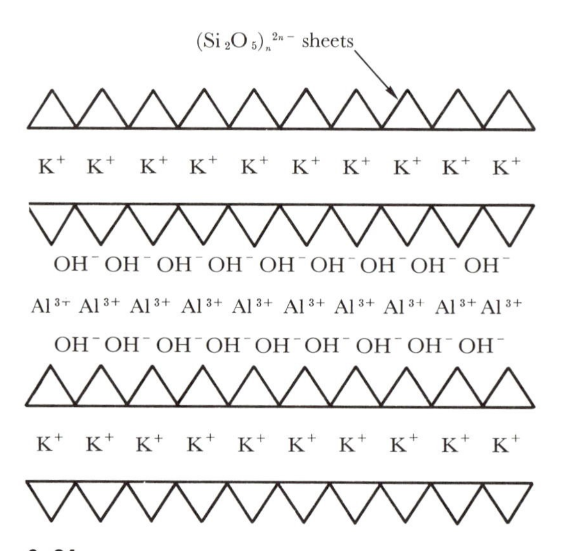

6–34
In the mica muscovite, $[K_2Al_4Si_6Al_2O_{20}(OH)_4]$, anionic sheets of silicate tetrahedra, as in Figure 6–33, alternate with layers of potassium ions and aluminum ions sandwiched between hydroxide ions. This layer structure gives mica its flaky cleavage properties.

Sheet structures. Continuous broadening of double-stranded silicate chains produces planar sheets of silicate structures (Figure 6–33). Talc, or soapstone, has this structure, in which none of the Si^{4+} is replaced with Al^{3+}. Therefore no additional cations between the sheets are required to balance charges. The silicate sheets in talc are held together primarily by van der Waals forces. Because of these weak forces the layers slide past one another relatively easily, and produce the slippery feel that is characteristic of talcum powder.

Mica resembles talc, but one quarter of the Si^{4+} in the tetrahedra is replaced by Al^{3+}. Thus an additional positive charge is required for each replacement to balance charges. Mica has the layer structure shown in Figure 6–34. The layers of cations (Al^{3+} serves as a cation between layers as well as a substitute in the silicate tetrahedra) hold the silicate sheets together electrostatically with much greater strength than in talc. Thus mica is not slippery to the touch and is not a good lubricant. However, it cleaves easily, thereby splitting into sheets parallel to the silicate layers. Little effort is required to flake off a chip of mica, but much more strength is needed to bend the flake across the middle and break it.

The clay minerals are silicates with sheet structures such as in mica. These layer structures have enormous "inner surfaces" and often can absorb large amounts of water and other substances between the silicate layers. This is why clay soils are such useful growth media for plants. This property also is why clays are used as beds for metal catalysts. The common catalyst platinum black is finely divided platinum metal obtained by precipitation from solution. The catalytic activity of platinum black is enhanced by the large amount of exposed metal surface. The same effect can be achieved by precipitating a metal to be used as a catalyst (Pt, Ni, or Co) onto clays. The metal atoms coat the interior walls of the silicate sheets, and the clay structure prohibits the metal from consolidating into a useless mass. J. D. Bernal has suggested that the first catalyzed reactions in the early stages of the evolution of life, before biological catalysts (enzymes) existed, may have occurred on the surfaces of clay minerals.

Three-dimensional networks. The three-dimensional silicate networks, in which all four oxygen atoms of SiO_4^{4-} are shared with other Si^{4+}, are typified by quartz, $(SiO_2)_n$. In crystalline quartz, all of the tetrahedral structures have Si^{4+} ions, but in other network minerals, up to half the Si^{4+} can be replaced with Al^{3+}. These minerals include the feldspars, with a typical empirical formula $KAlSi_3O_8$. Feldspars are nearly as hard as quartz. Basalt, which may be the material of the mantle of the earth, is a compact mineral related to feldspar. Granite, the chief component of the earth's crust, is a mixture of crystallites of mica, feldspar, and quartz.

Glasses are amorphous, disordered, noncrystalline aggregates with linked silicate chains of the sort depicted in Figure 6–35. Common soda-lime glass

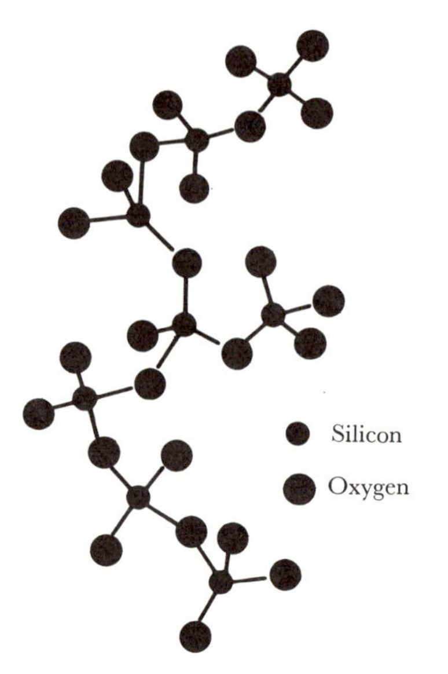

6–35
Glasses are amorphous, disordered chains of silicate tetrahedra mixed with metal carbonates or oxides such as Na_2CO_3 or $CaCO_3$.

is made with sand (quartz), limestone ($CaCO_3$), and sodium carbonate (Na_2CO_3) or sodium sulfate (Na_2SO_4), which are melted together and allowed to cool. Other glasses with special properties are made by using other metal carbonates and oxides. Pyrex glass has boron as well as silicon and some aluminum in its silicate framework. Glasses are not true solids, but are extremely viscous liquids.

The silicates illustrate all types of bonding in solids: covalent bonding between Si and O in the tetrahedra, van der Waals forces between silicate sheets in talc, ionic attractions between charged sheets and chains, as well as hydrogen bonds between water molecules and the silicate oxygens in clays. If we include nickel catalysts prepared on a clay support, metallic bonding is represented as well.

SUGGESTIONS FOR FURTHER READING

A. H. Cottrell, "The Nature of Metals," *Scientific American* (September, 1967).

W. A. Deer, R. A. Howie, and J. Zussman, *Rock Forming Minerals*, Vols. 1–5, Wiley, New York, 1962.

T. L. Hill, *Matter and Equilibrium*, Benjamin, Menlo Park, Calif., 1966.

W. J. Moore, *Seven Solid States*, Benjamin, Menlo Park, Calif., 1967.

N. Mott, "The Solid State," *Scientific American* (September, 1967).

H. Reiss, "Chemical Properties of Materials," *Scientific American* (September, 1967).

A. F. Wells, *Structural Inorganic Chemistry*, 3rd ed., Oxford Univ. Press, New York, 1962.

QUESTIONS AND PROBLEMS

1. What types of forces hold molecules together in crystals and liquids?
2. What effect do hydrogen bonds have on the boiling temperatures of liquids? Explain, and give an example.
3. Why are nonmetallic network solids usually quite hard?
4. What physical effect is responsible for the attraction in van der Waals interactions? What is responsible for the repulsion in such interactions? Compare the origin of attraction and repulsion in van der Waals interactions with that in ionic and covalent bonds.
5. How do we determine an experimental value for the van der Waals radius of hydrogen? From a theoretical viewpoint, what determines the van der Waals radius?
6. Draw a sketch of the way in which the repulsion part of the van der Waals interaction (Equation 6–1) varies with distance R between atomic centers. Draw a similar sketch for the attraction terms, Equation 6–2. Add these two curves in an approximate way and satisfy yourself that a potential curve such as Figure 6–10 is the result.
7. If van der Waals bonds are extremely weak, why are they discussed at all?
8. In the delocalized molecular orbital theory of metals, in what sense do we say that the entire piece of metal is a large molecule?
9. Why would beryllium be an insulator if the $2s$ and $2p$ molecular-orbital bands did not overlap?
10. What is the structural difference between metals, semiconductors, and insulators?
11. What effect do small amounts of boron or phosphorus have on the conducting properties of silicon?
12. How do hydrogen bonds participate in the structure of ice? What effects do they have on its properties?
13. How do we know that some hydrogen bonding in water persists in the liquid phase?
14. Provide a structural explanation for the fact that quartz is hard, asbestos fibrous and stringy, and mica platelike.
15. Why are clays useful in industrial catalysis?
16. It requires 5.2 eV, or 120 kcal mole^{-1}, to excite electrons in a diamond crystal from the valence band to the conduction band. What frequency of light is needed to bring about this excitation? What wavelength? What wave number? What part of the electromagnetic spectrum does this correspond to?
17. Using data given in this chapter, repeat Problem 16 for the semiconductors silicon and germanium.
18. Explain the trend in the melting temperatures of the following tetrahedral molecules: CF_4, 90°K; CCl_4, 250°K; CBr_4, 350°K; CI_4, 440°K.
19. Construct the potential energy curve for the Kr–Kr van der Waals interaction. How strong is the Kr–Kr van der Waals bond? Estimate the Kr–Kr bond distance in solid krypton.

20. The molecule RbBr is held together primarily by an ionic bond. The distance between Rb^+ and Br^- in the molecule is 2.945 Å. The closed electron shells of Rb^+ and Br^- both have the configuration of the noble gas Kr. From the energy curve constructed for Problem 19, estimate the van der Waals energy between Rb^+ and Br^-, assuming that the energy is the same as for a pair of Kr atoms separated by a distance of 2.945 Å. Is the repulsive part or the attractive part of the interaction dominant? How important is the van der Waals energy compared to the overall bond energy of 90 kcal mole^{-1} in RbBr? Examine the Kr–Kr van der Waals energy for distances of 2 Å and 1 Å and then explain what prevents Rb^+ and Br^- ions from approaching each other too closely in an ionic solid.
21. What type of solid will BF_3 and NF_3 molecules build? What kinds of intermolecular interactions are likely to be important in each case? Which compound should have the higher melting temperature?

Appendix

PHYSICAL CONSTANTS AND CONVERSION FACTORS USED IN TEXT[a]

Constants

Bohr radius	$a_0 = 0.529177$ Å
Electron mass	$m_e = 9.109558 \times 10^{-28}$ g
Electronic charge	$e = 4.80325 \times 10^{-10}$ esu
Pi	$\pi = 3.141593$
Planck's constant	$h = 6.626196 \times 10^{-27}$ erg sec
Rydberg constant (experimental value)	$R = 109{,}677.581$ cm^{-1}
Velocity of light	$c = 2.997925 \times 10^{10}$ cm sec^{-1}

Conversion factors

1 amu $= 1.660531 \times 10^{-24}$ g
1 erg $= 6.24145 \times 10^{11}$ eV
1 eV $= 8065.76$ cm^{-1}
1 eV $= 23.069$ kcal mole^{-1}

---

[a] We have used several sources in our effort to compile the latest values for physical constants, bond lengths, bond dissociation energies, ionization energies, and electron affinities. The principal references that we used are:

Physical constants, "Review of Particle Properties," *Rev. Mod. Phys.* **43,** S25 (1971);

Bond lengths and bond energies, Simone Bourcier (Redacteur en chef), *Constantes Sélectionnées Données Spectroscopiques Relatives aux Molécules Diatomiques*, Pergamon Press, Elmsford, N.Y., 1970; H. B. Gray, *Electrons and Chemical Bonding*, Benjamin, Menlo Park, Calif., 1965;

F–F bond energy, W. Stricker and L. Krauss, *Z. Naturforsch a. Dtsch.* **23,** 486 (1968);

Ionization energies, C. E. Moore, "Ionization Potentials and Ionization Limits Derived from the Analyses of Optical Spectra," NSRDS-NBS 34, National Bureau of Standards, Washington, D.C., 1970;

Electron affinities, H. B. Gray, *Electrons and Chemical Bonding*, Benjamin, Menlo Park, Calif., 1965; R. S. Berry, *Chem. Rev.* **69,** 533 (1969).

Index

ABCDEFGH79876543